Health, Technology and Society

Medicine, health care, and the wider social meaning and management of health are undergoing major changes. In part this reflects developments in science and technology, which enable new forms of diagnosis, treatment and delivery of health care. It also reflects changes in the locus of care and the social management of health. Locating technical developments in wider socio-economic and political processes, each book in the series discusses and critiques recent developments in health technologies in specific areas, drawing on a range of analyses provided by the social sciences. Some have a more theoretical focus, some a more applied focus but all draw on recent research by the authors. The series also looks toward the medium term in anticipating the likely configurations of health in advanced industrial society and does so comparatively, through exploring the globalization and internationalization of health.

Anna Molas

Taming Egg Donors

The Egg Donation Reproductive Market in Spain

palgrave
macmillan

Anna Molas
School of Social Sciences, Faculty of Arts
Monash University
Melbourne, VIC, Australia

ISSN 2946-3386 ISSN 2946-3378 (electronic)
Health, Technology and Society
ISBN 978-981-96-5102-3 ISBN 978-981-96-5103-0 (eBook)
https://doi.org/10.1007/978-981-96-5103-0

Cover illustration: Jahel Guerra

This Palgrave Macmillan imprint is published by the registered company Springer Nature Singapore Pte Ltd.
The registered company address is: 152 Beach Road, #21-01/04 Gateway East, Singapore 189721, Singapore

Series Editors' Preface

Medicine, health care, and the wider social meanings and management of health are continually in the process of change. While the 'birth of the clinic' heralded the process through which health and illness became increasingly subject to the surveillance of medicine, for example, surveillance has become more complex, sophisticated, and targeted - as seen in the search for 'precision medicine' and now 'precision public health'. Both surveillance and health itself emerge as more provisional, uncertain, and risk-laden as a consequence, and we might also ask what now constitutes 'the clinic', how meaningful a concept of a clinic ultimately is, and where else might we now find (or not find) healthcare spaces and interventions.

Ongoing developments in science and technology are helping to enable and propel new forms of diagnosis, treatment, and the delivery of health care. In many contexts, these innovations both reflect and further contribute to changes in the locus of care and burden of responsibility for health. Genetics, informatics, imaging - to name but a few - are redefining collective and individual understandings of the body, health, and disease. At the same time, long established and even ostensibly mundane technologies and techniques can generate ripples in local discourse and practices as ideas about the nature and focus of healthcare shift in response to global debates about, for instance, One Health and Planetary Health.

The very technologies that (re)define health are also the means through which the individualisation of health care can occur – through, for instance, digital health, diagnostic tests, and the commodification of restorative tissue. This individualisation of health is both culturally-derived and state-sponsored, as exemplified by the promotion of 'self-care'. These shifts are simultaneously welcomed and contested by professionals, patients, and wider publics. Hence they at once signal and instantiate wider societal ambivalences and divisions.

This Series explores these processes within and beyond the conventional domain of 'the clinic', and asks whether they amount to a qualitative shift in the social ordering and value of medicine and health. Locating technical use and developments in wider socio-economic and political processes, each book discusses and critiques the dynamics between health, technology, and society through a variety of specific cases, and drawing on a range of analyses provided by the social sciences.

The Series has already published more than thirty books that have explored many of these issues, drawing on novel, critical, and deeply informed research undertaken by their authors. In doing so, the books have shown how the boundaries between the three core dimensions that underpin the whole Series – health, technology, and society – are changing in fundamental ways.

Anna Molas' monograph, *Taming Egg Donors: The Egg Donation Reproductive Market in Spain*, explores these three dimensions through a focus on egg donors and donation in Spain's growing fertility industry. Based on multi-sited ethnographic fieldwork and careful, rigorous research and scholarship, Molas seeks to answer the question 'How are bodies stabilized for extraction purposes in the Spanish reproductive bio-economy?'. This stabilization is achieved, argues the author, through 'taming'; different methods of containment at the structural level of discourses and regulations, and within clinical practices, which allow the fragile project of egg donation to be effective. Through nuanced and engagingly presented analysis, Molas concludes that while smoothing the process for a successful egg donation cycle, such practices of taming also cause eggs donors to effectively 'disappear' from donation treatments while transforming their eggs into bio-objects ready to enter reproductive bioeconomy circuits. This highly original study of, and approach to, the technologies and relations fundamental to one of the largest players within the global reproductive bioeconomy forms an important contribution to the Health Technology and Society series and to scholarship on tissue economies, reproductive labour, and extractive industries in particular.

London, UK Rebecca Lynch
Edinburgh, UK Martyn Pickersgill

Acknowledgments

The years that took me to make this book represent one of the most intense, inspiring, and challenging periods of my life. The journey involved not only the writing of a manuscript, but also a relocation from Spain to Australia for my PhD and a three-year postdoc back in my hometown of Barcelona. I owe these lines to a large network of people who, across thousands of kilometers, supported me throughout this process.

First, I want to thank the participants in this research, egg donors and IVF practitioners who decided to take part in the project, who gave me their valuable time, and who had the humility to let their stories be interpreted by a third person.

I also want to thank Professor Andrea Whittaker who was the most excellent mentor I could have dreamed to have. I also want to express my gratitude to Professor Narelle Warren and Dr. Paul Mason, who were my second supervisors at different periods of my thesis. Thanks for your availability, advice, and more importantly, for your kindness throughout this journey.

I want to acknowledge the support of Professor Joan Bestard, to whom I owe more than anyone the opportunity to undertake this research in the first place. Thanks for, after being my MA supervisor, supporting me as I prepared a doctoral application. I will never forget those early days. I also want to thank Michal Nahman and Diana Marre for the advice and support provided as I converted my thesis into a book, as well as the two anonymous reviewers from Palgrave Macmillan, whose feedback was invaluable for deepening the main arguments of my manuscript. In this regard, I want to thank Marion Duval, publishing editor of the book series "Health, Technology and Society," for her support throughout the process. Finally, I want to thank Susan Frekko, who provided editing services for this book.

Throughout this time, a number of institutions, bodies, and research groups also supported me in different ways. I want to first thank Monash University, which funded me to undertake this research. Thanks also to the Spanish Ministry of Science and Innovation, which granted me the Juan de la Cierva Postdoctoral Fellowship (FJC2021-046533-I grant), that I undertook within the AFIN Research Group at the Autonomous University of Barcelona. I would like to also acknowledge the Research Group on Social and Cultural Geographies of the University of Bern and the Center for Science, Technology, Medicine, and Society of the University of California, Berkeley for hosting me as a visiting scholar and providing me with all the necessary support for moving on with my research. And most importantly, I want to thank the Independent Social Research Foundation (ISRF) for supporting the writing process with the First Book Grant 2023 award. Your help has truly been a game-changer for this project to come to fruition.

Thanks to my family, who found a way to always be close to me and to my project despite the distance, time difference, and language barriers: my parents, Marta and Prudenci, and my siblings, Mireia, Alba, Joan, and Maria. Thanks to my friend and colleague Laura Perler, with whom I share a strong intellectual and personal bond without which this thesis would probably look very different. To Berta, Albert, Sara, Marta, Anna, and Maria for being there, as always. And of course, thanks Ettore, Rudi, Sergio, Marco, Trang, Helen, and Don for the undisrupted laughter and adventures throughout my years in Australia and beyond.

Finally, to the person who stayed with me on both sides of the world, who embarked on this uncertain journey away from everyone and everything that we knew, with whom I have lived all the adventures of the past years and hopefully, the ones that are to come: thanks, Albert, for everything.

Contents

Abbreviations

AI	Artificial insemination
AMH	Anti-Mullerian hormone
ARTs	Assisted reproductive technologies
CBRC	Cross-border reproductive care
ECHR	European Court of Human Rights
ESCs	Embryonic stem cells
ESHRE	European Society of Human Reproduction and Embryology
GnRH	Gonadotropin-releasing hormone antagonists
ICSI	Intra-cytoplasmic sperm injection
INE	Instituto Nacional de Estadística (*National Statistics Institute*)
IUD	Intrauterine device
iPSC	Induced pluripotent stem cells
IVF	In vitro fertilization
CNRHA	Comisión Nacional de Reproducción Humana Asistida (*National Commission of Human Assisted Reproduction*)
OHSS	Ovarian hyper-stimulation syndrome
PGD	Preimplantation genetic diagnosis
ROPA	Reception of oocytes from the partner
SEF	Sociedad Española de Fertilidad (*Spanish Fertility Association*)
SIRHA	Sistema de Información de Reproducción Humana Asistida (*Information System of Human Assisted Reproduction*)
STS	Science and Technology Studies
TFR	Total fertility rate

List of Images

List of Illustrations

1

Introduction

Stabilizing Donors, Extracting Eggs

It is common knowledge that you get paid to donate eggs in Spain. I'm not sure when I heard about it for the first time. During my 20s, my social media profiles were full of advertisements from different clinics inviting me to become a donor. At my university's print shop, every print job was returned to the customer inside a free folder sponsored by a fertility clinic, which listed basic information about the clinic's egg donation program. This promotion strategy was still in place years later when I undertook my fieldwork. Once, I even received an email through the employment portal I subscribed to: "Donate eggs, REWARD!" said the subject line. The mix of surprise and anger that I felt about that message, which was purposely situated on the borders between an ad and a job offer, was the seed that motivated me to undertake this research.

The participants of this study had similar experiences to me. While some women got incessant advertisements on the Internet, others heard the story of an acquaintance or a friend of a friend who donated eggs, and from whom they learned, "It's like killing two birds with one stone, I earn money and I'm also helping someone who wants to have a child"

A. Molas, *Taming Egg Donors*, Health, Technology and Society,
https://doi.org/10.1007/978-981-96-5103-0_1

(Valentina, 21 years old). Indeed, these advertisements always pointed out the altruism of the donation. So, in the same way that everybody knows that you get paid to donate eggs in Spain, everybody knows that you are supposed to do it, at least to some extent, for altruistic reasons.

The aims of this research have changed slightly over time, particularly after conducting my fieldwork. My initial main purpose was to explore the experiences of egg donors, which is still a relatively neglected focus in the literature of the growing Spanish (and transnational) fertility industry. However, I quickly realized that centering the study on that question was insufficient for making sense of the entanglements that permeate the lives of the so-called egg donors. The first problem with this approach was, in fact, departing from the assumption that such thing as "egg donors" exist in Spain.

I am not referring here to the much-debated use of the term egg "donation," which presumes that the people who engage with this practice are motivated by the wish to charitably help other women become mothers (an issue on which I take a stance below). My surprise was rather to find out that my participants' ideas about egg donation varied significantly. Their literacy in the implications of the medical process, their understanding of the possible futures of their eggs, and their knowledge around the realities of the global circulation of ova and patients among Europe, to name some central aspects, varied greatly across the participants. Also, they rarely knew other women who had donated eggs. In this sense, I argue that this is an ethnographic *field without community*. I realized that such disarticulation was not a coincidence. Instead, it was one of the pillars on which the success of the egg donation industry rested.

Placing all of the focus of this book on their motivations and views would reveal donors' particular understandings and imaginaries about the process rather than any valorization of systems in place in fertility clinics or of the market dynamics that regulate the circulation of eggs, of which most of my participants were ignorant. This realization led me, while still placing the experiences of egg donors at the core, to focus on the mechanisms that bring women to complete extraction cycles.

The question that all the chapters of this book attempt to answer is: How are bodies stabilized for extraction purposes in the Spanish reproductive bioeconomy? This shift from motivations and experiences to practices allowed me to undertake a much more stimulating task, which in turn engendered more questions: What are the rationalities underlying women's decisions to undergo egg donation? How do these decisions relate to issues of information and risk perception inside and outside the clinic? What kind of labor is required by women engaging in a donation cycle? Under what conditions are donation cycles undertaken? How do clinics select women for extraction purposes? And finally, and most centrally, what kind of relations sustain this system? Indeed, this book is, above all, about a relationship: the relationship between egg donors and fertility clinics.

In examining this relationship, I show that egg donation is a fragile process. It is fragile in the sense that it needs to be carefully handled from the very beginning—the selection process of the donor—until the very end—when the extracted eggs are assigned futures. The process can fail at any point: The donor might not follow the rules properly; the drugs might not generate the expected effect; the risks cannot always be contained; the extracted eggs might not be mature enough; the children born might not meet the expected health standards. The environment of uncertainty, pressure, and high competition in which clinicians working on egg donation are usually embedded requires a diverse set of control techniques directed to minimize the associated risks.

The main argument of this book is that the whole egg donation system in Spain, both at its structural level (where socioeconomic factors as well as regulations converge) and at the micro level of clinical practices (from the personal relationships between clinicians and donors to the drugs they will be prescribed) are orchestrated to contain women's bodies and biographies in order to comply with clinics' extraction purposes. I contend that the process through which these two world orders—the one of fertility clinics and the one of donors, their bodies, and their eggs—are made ready to operate is a process of *taming*. My central aim is, therefore, to delineate the multilevel processes in place to ultimately achieve the stabilization of bodies and eggs, which is the condition for their entry into global bio-market circuits.

Taming Donors

The choice of the word *taming* to talk about the standardization process that takes place in different spheres of the egg donation process allows a new perspective on the relationships and practices that operate in fertility clinics. While the term evokes notions of control through an attempt to domesticate, the semantic charge goes beyond the framework of top-down discipline to account for other forms of ambivalent collaboration that include affective relations and mutuality. This term also allows me to enter into conversation with the scholarship on the political economy of reproduction by illuminating the labor involved in making eggs available for the global repromarket as well as its relational nature as a key site of value production. Throughout this book, I will use this term, taking inspiration from the study on animal-human affective intimacies by Aníbal Arregui (2020).

In his study, Arregui aims to disentangle the complex affective relationships between the endangered pink river-dolphin or *boto*—a popular character of Amazonian folklore—and the riverine dwellers of the lower Amazon region (Brazil). As he argues, humans and dolphins can show mutual hostility in some cases, while in others positive corporal engagements take place. The author concludes that highly ambivalent relationships happen in this interspecies encounter, which can be both seen as "wild" and "tame," and that, in either case, go much beyond the possibilities that the notion of "domestication" conceals. Situating his work in a long tradition of scholarship that acknowledges the agency and the potency of non-humans to shape human behavior (see Cassidy & Mullin, 2007; Haraway, 2008; Ingold, 2000; Tsing, 2012, 2015), he argues that "the analytical implications of distinguishing domesticatory processes from relations of taming still remain largely tacit in social sciences scholarship" (p. 824).

As I show in this book, clinicians perceive egg donors as entailing an animality and wildness, infused with gender, age, class, and racialized attributes, that they have to "tame" to ensure a successful egg donation cycle—and more desirable reproduction broadly. In this context donors are perceived as unstable biographies, fragile temporalities, and unpredictable bodies that might or might not align with clinical purposes. Or to put in Karen Barad's (1998) material feminist analytical terms, egg donors are perceived as matter always at risk of "kicking back" and complicating the egg donation process. Kalindi Vora (2016) describes "vital energy" as both biological and affective value that gets transmitted in reproductive third-party provision. Labeling the relationship with egg donors and clinics as a relation of taming highlights the methods in place to control, transform, and transfer vital energy in third-party provision. The context of this research, however, highlights that the dialectic that makes this extraction possible is underpinned by different ontologies of egg donors that exist in the fertility clinic. As I detail in the following chapters, egg donors are constructed as ambivalent subjects/objects that are "excessive" in different ways: They have excessive time, sexuality, reproductive capacity, (economic) instability, and emotions such as drama but also joy. These perceived excesses are constructed as extra, waste, or unnecessary, and this operation makes them available to be exploited, transformed, and mobilized.

The untamed dimension in which egg donors are infused also animates the historical colonial roots of human bodies as sites of extraction from "areas of life depletion to areas of life enrichment" (Vora, 2016, p. 6). As Alys Eve Weinbaum (2022) argues, even if not always embodied by racialized subjects, contemporary reproductive extraction and dispossession are possible due to a recalibration of racialized ideas of body parts as fungible commodities derived from the orchestration of enslaved women's reproduction as a means to reproduce the workforce, which makes reproductive labor an intrinsically racializing activity. Her historical account of the reproduction of enslaved people in the Atlantic as orchestrated through the implementation of managerial strategies resonates with the taming methods depicted throughout this book but complicates the presumed "top-down" nature of this relationship.

As I will show, taming emerges as a stratifying practice that reproduces existing inequalities in the sense that bodies perceived as having excessive vitality are selectively managed toward the reproduction of more desirable others. In this context, however, stratified reproduction acquires a temporal dimension: While many egg donors' reproduction is not actively constrained at the moment of donation—they often choose to donate before they wish to reproduce—discourses constructing their reproduction as undesirable play a central role. These discourses, shaped by intersecting imaginaries of age, class, gender, and race, but also by the lived experiences of egg donors themselves, foreground the basis upon which egg donation is made possible.

Following this line of inquiry, in this book I delve deeper into the possibilities of the concept of *taming* to shed light on the complex, bidirectional relations among egg donors, bodies, eggs, and clinicians. In this exploration, I argue that taming relationships, based on measures of control but also of care, concern, and affection, are an explicitly economic and value-producing axis within the Spanish repromarket. As Arregui claims, dolphins can be both "tame" and "wild beasts," which he does not see as static qualities of bodies, but instead *positional relations* that are negotiated. These positional relationships between donors and clinics take most of the attention of this book.

While the taming methods I outline recall in some ways the Foucauldian notion of "discipline" (explored particularly in *Discipline and Punish* 1979), the empirical findings led me to the decision not to make this concept central. For Foucault, discipline is the technique of power to manage the behavior of individuals in a society. It is based on a continuous surveillance of the subjects which is in part facilitated by social institutions such as the prison, the hospital, and the school. This book will show how egg donors are indeed under surveillance during the egg donation cycle. However, the process of taming is understood here as happening beyond the disciplining of donors as subjects. Instead, it also accounts for the containment of bodies and eggs, which are seen as having the potential to exert their agency, or "wildness" and make the egg donation process fail. Taming is therefore a notion that encapsulates disciplining measures but that also accounts for the more-than-human aspects in need of control and stabilization to achieve successful outcomes.

In the following chapters, I address the issue of the taming of egg donors through different perspectives: the taming of the discourses and understandings around the egg donation industry in Spain; taming through the labor of egg donation; taming through control, discipline, and affect; and the taming of the future of the egg cells.

On Nomenclature and Definitions

The Medical Process of Egg Donation

The European Society of Human Reproduction and Embryology (ESHRE) defines egg donation as a "fertility treatment directed to women unable to produce their own eggs or at high risk of transmitting a genetic disease. The donated eggs are fertilised with partner [or donor] sperm as in a conventional IVF treatment cycle, and one (or two) [and sometimes three (SEF, 2022)] transferred as an embryo for pregnancy" (ESHRE, 2017).

Egg donation consists of several steps. First, the ovaries of the donor are stimulated through self-administered hormonal drugs in the form of subcutaneous injections over 8–12 days (Image 1.1). The medication produces multiple mature eggs during one menstruation cycle. Once the

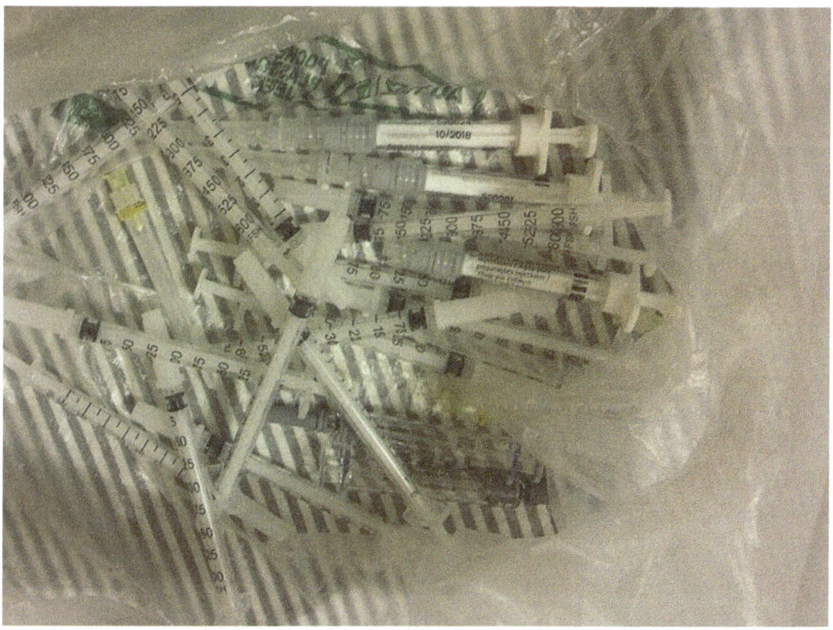

Image 1.1 Bag of used syringes of a donor halfway through the cycle

eggs are mature, they are retrieved from the donor through the surgical procedure of aspiration. The operation requires general anesthesia and usually lasts between 20 and 60 minutes. The number of eggs extracted in the puncture, their maturity, and their quality cannot be precisely predicted and are relative to the hormonal medication plan in place. However, the latest report from the Spanish Fertility Association (SEF) indicates that the average number of eggs obtained in Spain per extraction cycle in 2022 was 20.4.

Apart from the risks associated with anesthesia, egg donation is a procedure that carries some risks and side effects. As the SEF informs in their guidelines for the treatment of infertility (Matorras Weinig & Hernández, 2007), the most common is ovarian hyper-stimulation syndrome (OHSS). When occurring in women undergoing egg donation, it consists of an unexpected strong reaction to the stimulation treatment causing a cystic enlargement of the ovary. The SEF indicates that the frequency of the pathology is between 0.6% and 10% of women undergoing ovarian stimulation. It involves a significant growth of the ovaries, liquid accumulation in the abdomen or thorax, and alteration of renal and/or hepatic functions. In severe cases, it can be life-threatening, with the possibility of respiratory failure or coagulant alterations. It might require hospitalization and surgical treatment (Matorras Weinig & Hernández, 2007). Other risks and complications might include intolerance or side effects due to the medication, abdominoperineal infection, hemorrhage due to an accidental puncture of the blood vessels, abdominal pain, accidental puncture in another part of the body, and ovarian torsion.

While some studies have suggested a correlation between egg donation and certain kinds of cancer, long-term effects remain largely unknown due to the lack of longitudinal studies (Beeson & Lippman, 2006; Brody, 2017; Pearson, 2006; Schneider, 2008; Tober et al., 2021; Woodriff et al., 2014). The use of informed consent itself has been called into question by some who instead "call for more realistic explanations to egg donors about the lack of knowledge of long-term risks as well as more transparent informed consent documents" (Schneider et al., 2017, p. 480). Issues related to the information received by (intending) egg donors, the conditions in which it is delivered, and their risk perception are discussed in detail in Chap. 4.

Donors, Providers, or Sellers?

The terminology surrounding ova donation is debated (Beeson et al., 2015). While in most countries like Spain egg provision is embedded in the framework of altruistic donation, it seems obvious in most "repro-hubs" (Inhorn, 2015) that the term "donor" is a misnomer (Nahman, 2008) that hides both the economic interest and the reproductive labor that are usually involved in these transactions. Catherine Waldby and Melinda Cooper (2008) argue that the use of the word donation is "a way to expropriate donors and deny them rights over their bodily material" (p. 67), for which they coined the term "clinical labor," which reflects the precarious working conditions in a globalized bioeconomy (Cooper & Waldby, 2014). Following this line, Sara Lafuente-Funes (2019) argues that it is important to distinguish the work of egg donors or gestational surrogates from other assisted reproduction practices and rather talk about "transferences of reproductive capacity," which highlights specific power-relations operating in these exchanges by placing them as another step in the expanding feminized global care chains.

However, I observed that in the Spanish context, most women who underwent the process, regardless of their motivation and context, used the term "egg donation." Leaving this name behind in order to make a political statement as a researcher would also be abandoning the interesting and productive tensions that this term produces. These tensions are "generative nodes" (Haraway, 1991) that allow me to make sense of the participants' perspectives as well as the logics of the industry. In this book, therefore, I decided to respect the emic term "egg donation" used by all the participants, acknowledging (and embracing) its problematic potential.

Theoretical Discussion

Polyvocality in Egg Donation Studies

The relative dearth of qualitative studies that focus on the voices of egg donors around the world has been repeatedly pointed out by different researchers in the literature of assisted reproductive technologies (ARTs) (e.g., Lafuente-Funes, 2017; Molas, 2023; Nahman, 2018; Pollock, 2003; Rivas et al., 2018; Smietana et al., 2018). The perspectives of egg donors have been neglected traditionally in favor of the voices of fertility patients and clinicians or the expanding transnational dynamics of reproductive bioeconomies. Based on my own experiences in the field, I speculate that this relative absence is at least in part linked to a question of access. For example, during my fieldwork, I was not allowed to recruit egg donors through any of the clinics in which I interviewed clinicians or conducted observations. When clinics denied permission, they often highlighted the fear that donors would drop out of their donation journey if overly burdened by activities such as participating in my research. Unable to secure the help of the clinics in recruitment, I instead posted recruitment posters in several strategic sites of the city of Barcelona and outskirts, sent WhatsApp chains with information about the study, and contacted organizations related to motherhood issues that could disseminate information about my project through their social media accounts. Although in my case the outcomes were positive and several women contacted me to participate, the degree of uncertainty was very high, which is usually not tolerable for academic projects with tight timelines and budgets. As a result, there are relatively few studies that rely on egg donors' voices in Spain or elsewhere around the globe. Nonetheless, major ethnographic studies on egg donation have emerged over the past decades that shape our understandings of the practice, some using polyvocal approaches where both egg donors and clinicians are represented. While some of them are key references informing this book, others present some limitations that underscore the need for more ethnographic research on egg donation.

One of the first and most influential ethnographies on egg donors (and also on recipient women) is Monica Konrad's work *The Nameless Relations* (2006). Konrad's study, which was based in the UK and conducted prior to the ban on anonymous gamete donation in 2005, seeks to challenge the view that extracted eggs are automatically transformed into commodities in an anonymous egg donation system. Instead, the author sheds light on the different understandings of the gift created between altruistic egg donors and recipients by highlighting the kind of relationship that is "constituted in and through active not-knowing" (p. 243). According to her, even if not accomplishing the three rules of the Maussian gift—giving, receiving, and reciprocating (Mauss, 1969 [1925])—the transfer of ova between anonymous donors and recipients creates new relations and new kinds of relatedness. Establishing links with the findings of anthropologists in Melanesia in terms of exchanges and relationships—particularly those of Marilyn Strathern's *The Gender of the Gift* (1990)—Monica Konrad argues in favor of the existence of an anonymous sociality. Her study, however, draws from fieldwork conducted in mid-1990s. Since then, many aspects of egg donation have changed globally. Most notably, changes include the number of patients seeking these treatments, the number of women willing to donate their eggs (particularly in regions where a significant amount of money is given in exchange), and the number of countries that have progressively changed their laws regarding anonymity (including the UK).

A similar issue is observable in the ethnographic work on egg donors in Spain, where most studies are derived from fieldwork that took place in the 1990s and early 2000s (Alvarez, 2013; Bestard & Orobitg, 2009; Orobitg et al., 2013). A few scholars have more recently focused on the experiences of migrant donors in Spain and their participation in the industry (Nahman, 2018; Nahman & Weis, 2023); on their considerations in relation to the money received as well as the rationale that undergird their decision to undergo the process (Jociles Rubio, 2020; Rivas Rivas & Jociles Rubio, 2020), and their perceptions about the market changes brought about the routinization of freezing techniques (Lafuente-Funes et al., 2023).

An important body of work has explored different aspects of the Spanish reproductive bioeconomy. While these studies have not focused on egg donors' experiences, notable contributions such as those made by Sara Lafuente-Funes (Lafuente-Funes, 2021, 2023) have examined issues such as the ontologies of eggs in different settings (fertility clinics, labs, and universities), the role of egg donation in the functioning of fertility clinics and the discourses and practices that sustain it, and the role of vitrification techniques in new practices in fertility clinics' labs. Other authors have described the market design that ensures the availability of eggs and donors in Spain (Degli Esposti & Pavone, 2019); the similarities and differences of the Spanish and the US egg donation systems (Tober,

2024; Tober et al., 2023; Tober & Pavone, 2018); adoption, surrogacy and egg donation as forms of women's silenced reproductive work fueled by the framework of altruism and anonymity (Marre et al., 2018; Rivas et al., 2019); the contextual and legal causes for the growth of the Spanish bio-industry (Alkorta-Idiakez, 2006, 2010, 2021); and the links between egg donation and other types of feminized work that configure global care chains (Lafuente-Funes, 2019; Lafuente-Funes & Pérez Orozco, 2020). Some of these contributions are discussed further in the following chapters.

Among the ethnographic contributions in other parts of the world that use a polyvocal approach, the work of Rene Almeling in the United States deserves attention. In her book, *Sex Cells: The Medical Market for Eggs and Sperm* (2011), Almeling focuses on gamete donation agencies and donors. Through a comparative approach between sperm donation and egg donation arrangements, she concludes that gender matters in the organization of the two markets and the experiences of the participants. While egg donation is usually defined as an altruistic gift, sperm donation is rather framed as a job. As she contends, these discourses—promoted by clinics—reify the traditional ideas of women as dedicated care providers and men as breadwinners. Her work also provides an outstanding examination of how the biological discourses in clinics reproduce social constructions of gender, and how they have consequences for donors' understandings of the money they receive and their imaginary connection with the potential resulting children. Informed by Almeling's observations, my book shows how engaging with egg donation involves gendered clinical discourses as well as reproductive and emotional labor (Hochschild, 2000, 2003) that ultimately shape and tame egg donors' experiences, narratives, and subjectivities.

The contribution by Michal Nahman (2013) is of special relevance to this book. The author's focus on egg donors' everyday experiences beyond the donation itself as well as the situatedness of their lives in the historical, cultural, and socioeconomic context of a specific (peripheral) location make her work and findings very akin to my own ethnography. In *Extractions: An Ethnography of Reproductive Tourism* (2013), Nahman explores how the egg donation system in Israel co-produces national politics of identity in order to reproduce "Israeliness" in the context of the

Israel/Palestine conflict. From interviews with Romanian egg "sellers" (sic), the author highlights their agency in the extraction practices in which they engage. Nahman argues that they are "savvy participants" and "theorists of reproduction" who are nevertheless situated within the precarious neoliberal context of post-soviet Romania, where very limited options exist for women to devise a future.

Building on this foundational work, a growing number of scholars are examining egg donation arrangements in previously unstudied (or at least understudied) contexts such as Iran (Bonyad, 2024), Turkey (Mutlu, 2024), Ukraine (Vlasenko, 2024), Mexico (Perler & Schurr, 2020), South Africa (Moll, 2024; Namberger, 2019), Spain (Molas, 2023; Molas & Whittaker, 2023; Perler & Sánchez Pérez, 2024), and the United States (Tober, 2024). To foreground this emerging scholarship, Laura Perler and I co-edited the 2024 special issue "Reproductive labourers: centering the voices of egg donors and surrogates in the global bioeconomy" for *Medical Anthropology*. This issue aimed to trace the new directions of third-party reproduction studies that privileged the voices of providers. The articles in the issue highlight how ethnographic examinations of egg donation reveal the global fertility sector's capacity to adapt and exploit local frames, values, and inequalities. They also illuminate how participating in reproductive provision both produces vulnerabilities and capitalizes on pre-existing ones (Molas & Perler, 2024).

In this book, I use a polyvocal approach to explore the positionalities, objectives, and wishes that unfold respectively for egg donors and fertility clinics, as well as the dialectic process through which these become entangled and negotiated in a particular kind of relationship. By weaving together the detailed and in-depth accounts of both egg donors and clinicians, this book sheds new light on the relational dynamics that shape the egg donation repromarket. I argue that these dynamics can be understood as a process of taming, which at the same time constitutes an economic activity integral to value production within the Spanish egg donation repromarket. In doing so, the book examines the labor, temporalities, technologies, practices, and discourses that underpin the relations in this context. In the following sections, therefore, I present the main theoretical approaches that inform this book and how my contribution speaks to this rich body of literature.

Materializing Clinical Labor

The gendered division between productive and reproductive work continues to have a core function in the narratives and legal framework of egg provision in Spain, as in many other egg donation hubs. Although egg provision is embedded in the framework of altruistic donation in most systems worldwide, it has been argued that the term "donors" is a misnomer (Nahman, 2008) that obscures egg donors' socioeconomic contexts and the reproductive labor required of them. This perspective results in the invisibility of their work as it appears as emanating from a natural disposition of women toward caring for others, in the same way that reproductive labor more broadly was excluded from classical theories of value (Lafuente-Funes & Pérez Orozco, 2020; Marre et al., 2018; Rivas et al., 2019; Waldby & Cooper, 2008).

According to a long tradition of feminist scholarship, the Marxist framework is limited in its ability to account for the role of reproductive labor, as it focuses primarily on wage labor undertaken in spatially and temporally delimited spaces (Cooper & Waldby, 2014; Federici, 2012; Himmelweit, 1995; Vertommen & Barbagallo, 2022; Vora, 2016). As feminist scholars argue, instead, the reproductive work needed to maintain capitalism is ongoing, embodied (Fannin, 2020), emotional (Hochschild, 2000) and intimate (Boris & Salazar Parreñas, 2010) and consists of tasks such as food preparation, care for children and older adults, and sexual availability.

In their influential essay, Cooper and Waldby (2014) argue that the post-Fordist shift stabilized experimental and reproductive forms of value that were previously undertaken at the edges of the Fordist factory. Consequently, the advent of a deregulated and precarious labor market, along with technoscientific innovations, created the perfect conditions to convert the unpaid work of housewives and other (often gendered and racialized) subjects at the margins of the Fordist factory into different forms of "clinical labor." As the authors contend, this type of labor, which includes the provision of biological tissue, participation in clinical trials, and surrogacy, is central to biomedical innovation processes and post-industrial accumulation strategies.

Similarly, in her Marxist feminist analysis of egg donation in South Africa, Verena Namberger (2019) claims that a significant part of the labor demanded by egg donation is not acknowledged in the dominant narratives that see the body as "an organic whole infused with vitality" (p. 88), whose biovalue can only be extracted by the technological means of fertility clinics. As she contends, this approach undermines the material and immaterial labor that egg donors need to perform "from filling in the donor application form, through the hormonal stimulation and clinic visits to the retrieval of eggs and their fertilisation in the lab" (p. 88). These two studies highlight how the strategic undermining of the labor provided by egg donors is highly profitable for the expanding global bio-markets.

In this regard, the concept of "clinical labor" offers a very interesting lens to the study of egg donation, as it puts the focus on the financial profit or surplus value provided by egg donors' labor to the expanding bioeconomies, regardless of donors' own understandings of the nature of the donation itself or their motivations to undergo the process. In Chaps. 3 and 5, I locate this concept within the context of egg donors' lives in Spain. To do so, I describe, on the one hand, the intertwinements between women's labor lives and decision-making to become donors and, on the other, I identify and examine the different operations required for egg donors to make themselves "bioavailable" (Cohen, 2005).

In analyzing these operations in detail, I demonstrate, on the one hand, that the labor of donorship itself can be regarded as a process of subjectivation and taming, in which egg donors learn about the clinics' currency and their biocapital in the industry (Brown, 2015). On the other hand, I show how not only women's motivation to undergo the process, but also the means and the conditions in which they undertake it, are precariously articulated along the lines of race and class.

Temporalities, Risk, and Inequalities

Temporality is a central concept in the logics and practices of ARTs in general and egg donation in particular. A clear example of the importance of time is the fact that the exponential rise in the demand for IVF treatments using donated eggs in the last two decades is tightly connected to a delay of motherhood in Europe, as women struggle to balance childbearing with societal and work demands (see Chap. 2).

Temporal relationships in assisted reproduction are present in ethnographic studies that examine the rhythms of menstrual cycles, follicle maturation and embryo development as central concerns. These temporal aspects are intertwined with notions of reproductive aging, where patients race against their "biological clocks," or are defined by their age for treatment eligibility (Franklin, 2013) and also entangled with temporalities of marriage and sexual intimacy (Majumdar, 2021). These temporal dimensions are also noticeable in the liminal states of infertile women who are "not-yet-pregnant" (Sandelowski & de Lacey, 2002) and in the context of social egg freezing whereby women are encouraged to consider the possibility of future infertility in order to alter the timing of reproduction to align pregnancies with societal expectations regarding women's reproductive timelines (Baldwin et al., 2019; Inhorn et al., 2018; Waldby, 2019).

Katie Gentile (2015) explores the disconnection experienced by women undergoing fertility treatments, which she refers to as a state of "durationless present" produced by the repeated procedures, cycles, and uncertainties involved in the treatment process. Sarah Franklin (1997) captures this feeling by questioning how one can reach "the end of the story" when the causal and sequential order of events cannot be clearly established (pp. 12–13).

In their research on assisted reproduction in Switzerland, Bühler at al. (2015) delve into the complexities of temporal relationships within the ontological choreography of clinics. They argue that relational and historical time holds political significance for LGBTQI patients, whose pursuit to form families is influenced by past experiences of discrimination and aspirations for future innovative families facilitated by technology.

The concept of temporality in assisted reproduction has shed light on how emerging reproductive technologies, such as genetic screening and editing, freezing, bioengineering, and tissue transplants, place women within a new framework that emphasizes anticipation as a means to manage potential infertility or to regain opportunities for "missed" fertility

(Mansfield, 2017). These studies examine how women are encouraged to see themselves as future patients as well as political and economic actors, in narratives that promote anticipation as a form of care and risk management (Adams et al., 2009; Bach & Kroløkke, 2019; Martin, 2010). The advancement of freezing techniques also highlights the increasing commercially strategic use of these technologies which represent more efficient and profitable cycles (Lafuente-Funes, 2023; Waldby, 2019). These new rationalities are translated into "promissory capital" (Waldby & Mitchell, 2006) within the expanding fertility sector, which is transforming the management of fertility through a more speculative approach (van de Wiel, 2020).

Charis Thompson notes how assisted reproduction involves an "ontological choreography," which she defines as "the dynamic coordination of the technical, scientific, kinship, gender, emotional, legal, political, and financial aspects of ART clinics" (2005, p. 8). She describes the work of clinics in producing "a deftly balanced coming together of things that are generally considered parts of different ontological orders (part of nature, part of the self, part of society)" (2005, p. 8). In this book, and particularly in Chap. 4, I build upon her influential observation to consider explicitly how different temporalities are coordinated within fertility clinical practice and are used to tame egg donors in different ways. I contend that the anticipation and planning of future fertility co-exists and must be oriented in line with the temporalities of individuals who provide biomaterials or gestational services, which prompts us to question which bodies are directed to anticipate future risks such as infertility, and which bodies are directed to limit and circumscribe their temporality to serve the demands of the fertility industry (also in Molas & Whittaker, 2023).

Technologies of Control and Care

As this ethnography shows, the perception of egg donors as unstable women with complicated biographies permeates the narratives of clinicians in fertility clinics. In my fieldwork, while clinicians described some very responsible and motivated donors, they all acknowledged that most of their work with donors was laborious due to the women's unreliability. Departing from the stories of their struggles in dealing with them, I argue that the "reproductive biographies" (Perler & Schurr, 2020) of intending egg donors are perceived as in need of taming to enter and successfully deal with the egg donation process (see Molas & Perler, 2020). As I discuss particularly in Chaps. 6 and 7, the selection and taming techniques in fertility clinic to manage, on the one hand, their behavior, and on the other, their bodies and eggs, are directed to control them as well as to build relations of reciprocity that ultimately lead them to collaborate with the clinics. I examine the selection and management practices in fertility clinics to delineate how egg donors are tamed through techniques in which the boundaries between control and care get blurred.

The finding that third-party reproductive laborers are subject to disciplinary practices and required to perform emotional labor is not new in reproduction studies. It is actually a logical consequence that is already implied in Melinda Cooper and Catherine Waldby's formulation of "clinical labor" (2014) that has its roots in forms of unpaid gendered work that supposedly emanate from the natural caring drives of women, whose intimacy becomes increasingly commodified in the advent of global capitalism. Some scholars have focused on the mobilization of feelings that result from the clinical discourse of altruism, which situate participants in complicated dichotomies of good women versus selfish, financially driven or prostitute women (Almeling, 2011; Hovav, 2019; Pande, 2014; Rudrappa & Collins, 2015). Others have examined how the emotional labor of reproductive workers is disciplined to disentangle affective ties toward resulting children (Siegl, 2018; Teman, 2010). And finally, scholars have described the management of the relationships (or the absence of thereof) between donors and intending couples (Weis, 2021; Whittaker, 2018).

In her examination of surrogacy practices in Russia and Ukraine, Veronika Siegl (2018) draws on the concept of "technologies of the self" from Foucault (1988) and "emotional labor" from Hochschild (2003) to formulate what she calls "technologies of alignment." With this term, she refers to the process that "surrogate workers fulfil and are expected to fulfil, in order to find the right balance between care and distance in relation to the child they carry, and not to 'fall victim' to their hormonal bodies" (p. 64). In a similar fashion, Christina Weis (2017) observes that, in Russia, surrogates subjugate themselves to the "rules of the game" temporarily to achieve their own goals:

> The majority of surrogacy workers manifested their agency and pursued their goals of earning a big sum of money in the shortest possible time not by resisting the structure within which they played their "serious games," but by adapting to and temporarily accommodating the imposed rules. More often than changing the rules of the game, they reproduced them. (p. 217)

Nevertheless, while these discipline mechanisms are usually described as happening at the discursive level, in this book, I focus on control in terms of care and affective relations that are materialized through practices, legal documents, technologies, and drugs and that generate intimate labor for both reproductive laborers and clinicians. According to Boris and Salazar Parreñas (2010), "intimate labor":

> encompasses a range of activities, including bodily and household upkeep, personal and family maintenance, and sexual contact or liaison. It entails touch, whether of children or customers; bodily or emotional closeness or personal familiarity, such as sexual intercourse and bathing another; or close observation of another and knowledge of personal information, such as watching elderly people or advising trainees. Such work occurs in homes, hospitals, hotels, streets, and other public as well as private locations. It exists along a continuum of service and caring labor, from high-end nursing to low-end housekeeping, and includes sex, domestic, and care work. (p. 2)

In other words, intimate labor is about closeness with another person's body, sensitive personal information, or living spaces such as the household. As I show throughout this book, clinics become spaces of intimate relationships between donors and clinicians, who share aims, desires, and usually also struggles, that, despite their different positions, they negotiate together. Highlighting how descriptions of "intimate labor" have focused on its immateriality, I agree with Ariel Ducey (2010) when she argues that technology has been written "out of the story of caring (and affective) labor" (p. 19). According to her,

> [C]are can be extended to things, just as people can be treated like tools or commodities. Although objects are not usually considered the focus of concern in institutions and relationships that provide care, the objects that figure in such institutions can arguably become the source of intimate and caring relationships in their own right or the basis of unfolding awareness and what Bruno Latour would call increasingly articulate bodies. (p. 20)

Ariel Ducey refers in this quotation to Bruno Latour's (2004) proposal to understand bodies beyond the dualism between physiology and phenomenology. As he contends, having a body is a process of articulation in which not only biological, but also artificial and material components take part, and through them we ultimately learn to be affected. Ducey's (2010) reframing of the understanding of care relations to include non-human elements is crucial in my descriptions of the relationships between donors and clinics, where discourses but also practices, technologies, and substances intervene to shape, mediate, and contain the interactions to achieve an outcome that is deemed successful.

Understanding "intimate labor" as encompassing this non-human dimension, I rely on Andrea Whittaker's (2018) proposal to view surrogacy and egg donation as forms of "intimate *and* clinical labor," which "allows for a more nuanced approach that recognizes the significance of the social, cultural, economic, and political structures that shape it" (p. 52). Through this focus on intimate and clinical labor, this ethnography not only accounts for the external conditions through which donors enter the Spanish (transnational) reproductive bioeconomy, but also attends to the dynamic relationships of taming that align and convert certain reproductive biographies into valuable bio-objects (Holmberg et al., 2011) for the global bio-market circuits.

Some work on reproductive laborers has moved in this direction, but the only published ethnographies are of surrogacy arrangements rather than egg donation. Amrita Pande (2014), for example, explores how surrogacy agencies socialize Indian surrogates through the different steps of the process—from recruitment and contract to their stay in surrogacy hotels, childbirth, and post-partum—in an aim to convert them into the perfect "mother-worker." This process entails discipline and control techniques as well as affect and care toward surrogates. Sharmila Rudrappa (2012) targets this issue in her ethnography on surrogacy in India, where she accounts for relations of reciprocity and power that operate in the agencies and hostels. Both authors highlight how forms of coercive control and discipline are deployed alongside methods of mutual care between staff and reproductive laborers, who participate actively in this relationship, contesting, negotiating, and sometimes reifying clinics' discourses in a way that makes it difficult to trace clear lines between coercion, reciprocity, willingness, and love. As Amrita Pande (2014) puts it in her ethnography,

> Surrogates resist the construction of surrogates as "disposable and dirty workers" and as "disposable mothers." In sharp contrast to the surrogate-prostitute analogy used by the clinic and the media, the surrogates construct themselves as moral and needy mothers. While on the one hand, these alternative constructions of surrogacy by the surrogates affirm their dignity and sense of self-worth, on the other hand, they replicate certain gender hierarchies. Ironically, while the focus of this study has been on surrogacy as labor, most surrogates do not speak of surrogacy as paid work or work that they have chosen. It is almost as if the surrogates do not resist the image of women as selfless dutiful mothers whose primary role is to serve the family. (p. 168)

My book fills this gap in the literature on egg donation by focusing on how control, care relations, and intimate labor materialize in the clinic as yet another form of taming egg donors, who enter the highly emotional space of the fertility clinic. In this regard, I focus not only on the emotional labor demanded of egg donors, but also on the intimate labor performed by both donors and clinicians, especially by administrative staff, nurses, and psychologists through practices, documents, technologies, and substances. These material and immaterial forms of taming bring to life the interrelations of technoscientific products, state biopolitics, biocapital growth, and medical practice.

Stabilizing Genetics and Ethnoracial Pasts

Another aspect that needs to be tamed for the egg donation cycle to succeed is the egg cells themselves. Gametes outside bodies are contested matter. They entail ethical debates about their extractions and future use, and they do not stand either as live matter or as simple objects. Through technological innovations, frozen gametes can be suspended in time, making it possible to reproduce individuals at different moments (Waldby, 2014, 2019); they create dilemmas at the level of storage time and the possibility of posthumous reproduction (Kroløkke & Adrian, 2013); and finally, their existence isolated from the bodies, only possible through technoscientific interventions and egg donors' material and immaterial labor (Namberger, 2019), challenges the boundaries of categories, making them "matter out of place" in the sense of Mary Douglas (Douglas, 1966; see Holmberg et al., 2011). Their ambivalent condition, therefore, will need taming measures to control the dangers and ambiguities it entails. In this regard, egg cells are the perfect example of what Tora Holmberg et al. (2011), and Ingrid Metzler and Andrew Webster (2011) call "bio-objects." As these authors put it, "bio-objects" are "contested socio-technical objects" (Holmberg et al., 2011, p. 2) that

> tend to disrupt the conventional boundaries and identities of biological forms and categories, such as the boundaries between humans and animals or between the natural and artificial, sitting ambiguously in between those entities that we tend to conceptualize as human subjects and as non-human objects instead, sometimes troubling or even unsettling this very distinction. (Metzler & Webster, 2011, p. 649)

Stabilization is crucial for eggs to enter the circuits of biocapital, as their unsettling nature needs to be converted into promises of particular futures for intending parents. On the one hand, eggs must come with ethical assurances, a process that is accomplished with the informed consent process with donors. On the other, eggs must be assessed in terms of quality through different medical screenings of the donors, with a special focus on genetic diseases transmissible to the offspring. Finally, they must be properly categorized in terms of ethnoracial parameters—or phenotypical features, as the clinics usually put it—in order to make "proper" matchings with intending parents.

This process of conversion is ongoing and does not have a clear end. This is both because the attempts to classify eggs into stable categories are always contested by the different actors involved and because the categories themselves mutate over time with the change of paradigms, imaginaries, political regimes, and scientific novelties. This book shows how these processes are never easy to undertake and generate discomfort among clinicians, who are required to make complicated decisions that animate taboo imaginaries of eugenics and race selection. I also show how, even if thoroughly done, these selection processes cannot provide absolute certainty about the resulting children, which leaves a sense of risk.

Genetic standardization for egg donation is achieved through genetic tests of intending egg donors. Spanish legislation has little to say on which conditions should be screened for and which ones should lead a donor to be rejected. The law merely states that the

> [egg donor's] psychophysical state will have to accomplish the demands of a mandatory protocol of donors' screening which will include their phenotypical and psychological characteristics, as well as the clinical conditions and the needed analytical determinations to prove, according to the state of the art and the existing techniques at the moment of their realization, that donors do not suffer from genetic, hereditary, or infectious diseases transmissible to the offspring. (Law 14/2016, pp. 5–6)[1]

The now-routinized genetic carrier screening in Spanish fertility clinics has provided interesting new logics of selection. As Arribas-Ayllon (2011) states, "Genetic testing not only produces knowledge that we are 'fettered to,'" but also,

> new genetic knowledge is continuously negotiated and contested. The laboratory procedure of a "genetic test" is only one dimension that has changed our knowledge of health and disease or our expectations of prevention and treatment. We argue that responsibility for the calculation and management of genetic risk is a situated activity accomplished by local actors. (p. 4)

[1] Original text in Spanish. Fragment translated to English by the author of this book.

Investigating these dimensions of the genetic knowledge and risk as situated and contested, this book explores the rationalities of clinicians around genetic carrier screening. As I outline below, clinicians struggle to discard genetic pasts as the process itself gives rise to notions of eugenics that recall the consequences that such discourse had in Europe in the twentieth century (Arribas-Ayllon, 2011; Wade, 2012; Wahlberg & Gammeltoft, 2018).

However, market dynamics—the forces that move and direct the present and future of genetic science—place more value on "clean(er) donors": those who carry no-prevalent genetic diseases and who are, therefore, more easily marketable within different countries' regulations. This exploration makes clear how, despite cultural and medical discourses around how genetic pasts should be selected, market dynamics have an important role at the practical level. These dynamics highlight again the logics of "liberal eugenics" as "the unintended consequence of numerous individual choices made under the guise of technical and economic efficiency" (Arribas-Ayllon, 2011, p. 54).

Finally, I examine how the knowledge provided by genetic carrier screening impacts the subjectivities of egg donors who are labeled "genetically at risk" (Novas & Rose, 2000). As Arribas-Ayllon (2011) puts it, I look at

> the ways in which the testing of inherited risk enjoins new forms of social, individual and professional responsibility, and by extension, provides new resources for blame. Families and professionals, entrepreneurs and consumers, and "the public" more generally, are being drawn into discussions and decisions, which are changing the ways in which we think about ourselves and our relations with others, about future risks and how these can be minimised, if not avoided altogether. (p. 2)

Indeed, as has been argued with other kinds of genetic testing such as preimplantation genetic diagnosis (e.g., Franklin, 2006; Kathryn & Clare, 2010; Pavone, 2017; Pavone & Arias, 2011; Pavone & Lafuente Funes, 2018) or DNA ancestry testing (e.g., Halovic, 2017; Lacaze et al., 2017; Mason, 2017), the knowledge produced through genetic carrier testing has implications for both the lives of the tested (egg donors) and the lives of consumers (recipient couples). I discuss some of these consequences in the context of egg donation in Spain in Chap. 7.

The Spanish law on assisted reproduction states that clinics have to provide intending mothers with gamete donors who are as similar as possible to their phenotype (Law 14/2006). The function of mandatory phenotypical matching is to produce "plausible" babies (Thompson, 2009), or as Peter Wade puts it, children with "race-kinship congruity" (Wade, 2012); that is, babies with the ability to "pass" as biological offspring of the intending couple (Bergmann, 2012, 2014). This might be related to the aim of hiding the nature of the conception from whomever the parents wish, as well as avoiding additional stigma for the children of LGBTQI parents, who already face discrimination at different levels (Smietana & Twine, 2022). However, deciding in matters of resemblance is not such an easy task as the law on assisted reproduction seems to imply. As I unpack in Chap. 7, phenotypical selection is also a contentious matter for clinicians who, despite their attempt to avoid "racist" dynamics, reanimate racial categories in the matching practices themselves.

As Thompson (2009) asserts when discussing the context of egg donation in the US, "perceiving skin tone always involves its intersection with these other attributes and their wider meanings and histories. This means that skin color is an index of legibility whose chromatic properties are deeply relative" (p. 132). Race is, in this context, enacted rather than discovered in the negotiations between patients and doctors and in the selection of egg donors. Along the same lines, Laura Mamo (2005) highlights how relatedness is never only based on bio-genetic ties and social ties. According to her, "Biology and sociality are mobilized to create shared ancestry and the similarities and histories that embodies" (p. 254), something that Charlotte Kroløkke named "affective assemblages" (2014). In a similar vein, Jaya Keaney (2023) examines race as a reproductive technology in which "both the biological 'reality' of race and our racialized identifications are continually in an open-ended process of articulation" (p. 21). These authors highlight the inevitable role of ethnoracial imaginaries in the presumably objective selection of phenotypes. The attempt to tame ethnoracial pasts into stable categories serves "as a specific context for the fabrication of race" (Bergmann, 2015, p. 239).

This book focuses on the decision-making process of clinicians when taming egg donors' ethnoracial pasts or, in other words, when stabilizing their phenotypical characteristics to match with intending parents' or to freeze their eggs in the bank for later use. I explore how these practices enact a particular definition of race that is presented as a consequence of a society with racial prejudices combined with market dynamics. The mandatory practice of phenotypical matching places the clinicians in the challenging situation of trying to avoid racist dynamics while at the same time, reproducing racial tensions through the tailoring or "curation" (Moll, 2019) of the phenotypical match.

In sum, in this book I shed light not only on egg donors and clinicians' positionalities and desires, but also on how they are made to work in a particular kind of relationship. By asking how bodies and eggs are stabilized to serve the Spanish egg donation bioeconomy, I reveal the relational, affective, and material practices in place to secure eggs for the global reproductive market. This process of stabilization is what I call a process of taming. Throughout the chapters of this ethnography, I show how this relationship is situated within broader structural forces such as socioeconomic, class, gender, and racial legacies and inequalities as well as regulatory frameworks, public discourses, and reproduction politics.

Methodology and Epistemology

I used ethnography as the learning approach for data collection, basing my research interpretations on the practices of my participants and the meaning they attributed to them (Schensul et al., 1999). In particular, I relied on the conceptual approach of "multi-sited ethnography" (Marcus, 1995, 1998), which challenges the conventional conception of single-site ethnographic research as the main means to achieve in-depth knowledge through intimate face-to-face encounters with communities.

As Marcus (1995, 1998) notes in his germinal essay, multi-sited ethnography has been especially prevalent among cultural studies of science and technology, in which, due to the nature of their research interests, single-sited fieldwork was not a satisfying option. The approach of researchers such as Bruno Latour and Donna Haraway of "following the thing" in a metaphorical and material sense demands a new way to understand both the sites, the subject/object of study, and the researcher. Multi-sited ethnography allows for these choreographies as it is "designed around chains, paths, threads, conjunctions, or juxtapositions of locations in which the ethnographer establishes some form of literal, physical presence, with an explicit, posited logic of association or connection among sites" (Marcus, 1995, p. 105).

Multi-sited ethnography is not a way to be closer to the "reality" of the issue under study. Different sites are not intended here to provide a more complete vision of the picture, but instead to allow me to produce a more complex and interconnected one. Far from aiming for "objective" representation, by taking several standpoints, I intend to produce knowledge that is deeply contextual and embodied. Therefore, I engage with the proposal of Donna Haraway of "situated knowledges" (1991), which accounts for the production and relevance of the embodiment of the researcher as well as the historical and cultural production of the "objects" of study, that she labels as "material-semiotic actors" (p. 200). Precisely because of this complex genealogy that configured the boundaries of such objects/subjects, Haraway attributes to them a great deal of agency. As she claims, "accounts of a 'real' world do not, then, depend on a logic of 'discovery,' but on a power-charged social relation of 'conversation'" (p. 198). This is the way I conceive of the present research, as a situated conversation between myself as an embodied sociocultural subject and the material-semiotic actors that came to be the objects/subjects of this research in a series of produced sites.

A number of contingent issues led my research project to take place primarily in the city of Barcelona, Spain. On the one hand, Catalonia is a significant geographical place for the fertility industry, as it is the most popular destination for assisted reproduction across Spain (SEF, 2022). On the other hand, as I was a local and had based my previous research there, my knowledge about the history, the social context, and the egg donation assemblage made it easier to arrange and make sense of my fieldwork experience within the time limits of this project. However, Barcelona was only the point of departure and the physical space in which the observations took place. As stated before, my research was multi-sited in practice, as I not only entered the realm of different fertility clinics, but also followed my key participants through the different spaces of their lives, conducting interviews in their hometowns, through the Internet or virtually accompanying them in their lives through fragments of diaries, video-recordings, or voice messages they shared with me.

The multi-sited approach of this book materialized with the selection of a diverse number of actors and methods that revealed different aspects of the Spanish reproductive bioeconomy. The polyvocal narratives encountered in the various field sites provided me not only with the particular insights of my participants but also with the socio-technical imaginaries operating in fertility clinics and among egg donors. The following illustration (Illustration 1.1) offers a summary of the methods used and the number of participants of this research.

Semi-structured interviews represented one of the most important sources of information for this research, and through them I talked to two broad groups of participants: (intending) egg donors and professionals from fertility clinics.

I interviewed a total of 25 women, 22 of whom did at least one cycle and three of whom intended to donate eggs but were rejected from the program, and 25 current or former clinicians of six different fertility clinics. Among this group, there were gynecologists, embryologists, lab directors, embryologists, psychologists, and administrative and international patient coordination staff.

The interviews were held in a convenient place for the participant, which usually was a cafe near their homes or places of study or work. Due to the long distance for some of the participants, five interviews were held on Skype. I had two or more follow-up interviews with nine of the participants. In addition, three of them also participated in a focus group.

Finally, I conducted observations at a fertility clinic where I could be present in lab routines, at the reception desk, in the consultation room with the gynecologist, and in the operating room during four egg extraction surgeries. Although access was a very difficult aspect of my fieldwork and the time that I was permitted to perform observations was limited (all observations happened in the course of one month)—the quality of information that these experiences provided was very important to my overall understanding of egg donation practices.

METHODS	In-depth interviews Focus group discussion Negotiated interactive observation
EGG DONORS	22 initial in-depth interviews (demographic data in Appendix) 9 follow-up in-depth interviews 1 focus group discussion with 3 egg donors
REJECTED DONOR APPLICANTS	3 initial in-depth interviews (demographic data in Appendix) 1 follow-up in-depth interview
FERTILITY CLINICS' STAFF	22 in-depth interviews (with 25 professionals, as two were group interviews) Participants across 6 different clinics
STAFF BY ROLES	7 biologists 6 gynecologists 6 international patient coordinators 3 nurses 2 psychologists 1 administrative staff person

Illustration 1.1 Methods and participants

Notes on Language, Translation, and Ethics

My conversations with participants happened either in Catalan or Spanish, my native languages. I transcribed them in the original language and later I translated the selected quotations to English. Also, the field diary was originally written in Catalan and later I translated the specific fragments that appear in the book. This book also contains references to official documents (laws, informed consent forms, and reports) and academic papers that were originally written in Spanish. These have been translated by me and are followed by a footnote indicating the original language of the document. As the reader will see, in some parts I maintain original terms in Catalan or Spanish, putting their meaning in English in brackets. I did this especially when the translations to English did not capture the full meaning of the term in Catalan or Spanish, which was usually when the participants cursed or used idioms.

All participants in this research, whether involved in interviews or present during the observations, were provided with detailed information about the study. This information was typically given in the form of a written explanatory statement, followed by an oral explanation and the signing of an informed consent form. In the few cases where providing physical documents was not possible during observation—primarily due to time constraints of the participants—the information was delivered orally by me prior to data collection. During the observations, I took notes whenever it was possible without disrupting the work or interactions of professionals and patients. The real names of participants were never recorded in the field notes and have been substituted with pseudonyms to ensure confidentiality.

Chapter Outline

This book is comprised of nine chapters. In this introductory chapter, I have outlined the aims and goals of the book as well as the theoretical framework and the methodology used. Chapter 2, "Spain's Transnational Fertility Industry," concerns the context of the research, where I provide key information regarding the egg donation industry in Spain as well as aspects of its political and socioeconomic situation. After this, the book has five findings chapters, a conclusions chapter, and an epilogue.

In Chap. 3, "Securing a Future," I discuss the intimate relationship between egg donation and the precarious labor market in Spain. Drawing on participants' stories, I argue that a framework that uses the dichotomy of altruism/economic motivation is not useful for providing an understanding of the ways egg donors make sense of their actions. I contend that while, for most, the altruism of the donation has a function in the articulation of their stories and framing their narratives, economic compensation appears as the foremost driver for undergoing a donation. The dichotomous framework of altruism/economic interest has effects beyond discursive practices and materializes in their stories as feelings of ambivalence, guilt, and shame. It highlights the gendered expectations for women to be altruistic rather than financially motivated, which is also promoted by the clinics' discourse. In examining the stories of my participants with egg donation, I argue that most of them draw on a common source: the precarious labor market in Spain in its different ramifications.

In Chap. 4, "Tempting luck," I examine how structural factors and clinical practices in Spain intersect to circumscribe egg donors within a restricted temporality. I focus on the relationship between the quality of the information provided, informed consent arrangements, and the anonymity framework of egg donation, to describe how these tame donors' imaginaries by erasing both the long-term risks and the future distribution and use of their eggs. Drawing on the experiences of several participants, I highlight how egg donors' young age, implicit trust in doctors, scant cultural capital, and the anonymous framework play a role in their limited capacity to anticipate and manage risks. By using the notions of "temporality" and "risk anticipation," I analyze both the structural and individual arrangements that ultimately tame egg donors, separating them from the value of their contribution and precluding claims for retribution and risks, while enabling the growing fertility industry to exert ownership over the extracted eggs.

In Chap. 5, "Laboring Donors," I discuss the different operations required for intending egg donors to make themselves bioavailable, asserting that each of them configures the unacknowledged labor of donorship. I examine donors' process of choosing a clinic, arguing that while participants embarked on this selection process to assert agency and control, they eventually realized that they might not be able to choose, grasping in this way their biocapital in the fertility industry. I also explore the practical and emotional demands of the egg donation process such as self-medication, work arrangements, and other life commitments that were often precariously balanced with egg donation. Finally, I focus on the pain and discomfort egg donors endured and how, contrary to the discourse of fertility clinics, women performed significant amounts of emotional and physical labor that needed to be managed before, during, and after the extraction. In this chapter, therefore, I illuminate the specific kind of reproductive labor of egg donors, including both its material and immaterial spheres. I argue that these efforts extend beyond the act of donation itself and that they subjectivize and tame women in particular ways.

In Chap. 6, "Containing Behavior," I propose an ontological duality of egg donors enacted in fertility clinics. Focusing on the practices and discourses in this setting and drawing from Annemarie Mol's conceptualization of multiplicity, I distinguish between the biographical subject and the body object. This chapter focuses on the delimitation, selection, and taming practices of the biographical subject. I describe how the donor was enacted as a subject with agency that would have to be contained throughout the donation process in order for the cycle to be completed. In other words, clinics aimed to select a responsible biographical subject who obeyed professionals' instructions throughout the process in order to make the body object accessible for the clinics' extraction purposes.

In Chap. 7, "Labeling Ova," I delve deeper into the argument about the ontological duality of egg donors in fertility clinics started in Chap. 6, focusing on the delimitation, selection, and taming practices of the donor. First, I explore the way in which genetic carrier screening was managed in fertility clinics. I argue that the law stating that clinics should screen according to the current state of science and the techniques available is problematic, as it gives license and justification to a highly profitable relationship between clinics and genetic labs. Second, I explore the consequences that the results of genetic carrier screening had for egg donors, explaining how undergoing this test can create new subjectivities and responsibilities when prevalent genetic diseases are found. Third, I explore how phenotypical matching was navigated in the clinic such that the fertility clinic became a site of negotiation, contestation, and reproduction of racial tensions.

In the Conclusions, Chap. 8, I argue that, taken together, these chapters provide a picture of how women's biographies and reproductive capabilities were set to effectively guarantee successful egg retrievals that could promise specific genetic and ethnoracial futures and qualities. To achieve this, taming methods were present at the discursive level, where a particular framing of egg donation as altruistic and anonymous generated certain kinds of understandings among donors; at the level of information delivery in the clinic with the obscuring of the transnational market dynamics in which egg donation was embedded; at the level of the selection and management of egg donors' biographical profiles, where different control and caring techniques were used to keep them in line; and at the level of their bodily reproductive capabilities, where their genetic and phenotypical information was precariously stabilized in objectifiable categories. This process ultimately undermined the participation and labor of egg donors and made them disappear from egg donation treatments, transforming their eggs into bio-objects ready to enter the reproductive bioeconomy circuits.

Finally, in the Epilogue, Chap. 9, I discuss clinicians' imaginaries about the future. Mindful of the fragility of the egg donation process and the challenges of stabilizing eggs and women, clinicians were impatiently waiting for the time in which technoscientific innovation would effectively remove egg donors from the equation of fertility treatments. I delve into the ambivalent relationship that clinics had with donors, who were seen both as necessary and troublesome. Interestingly, the debates related to the problems of egg donation—such as the poor information donors received or the precariousness that underpinned their availability—were seen as somehow irrelevant because they were presumed to be temporary.

In order to provide an overview of the context of this research, in the following chapter, I first describe some key aspects of the Spanish legislation on assisted reproduction and the functioning of the Spanish fertility industry.

References

Adams, V., Murphy, M., & Clarke, A. (2009). Anticipation: Technoscience, life, affect, temporality. *Subjectivity, 28*(1), 246–265. https://doi.org/10.1057/sub.2009.18

Alkorta-Idiakez, I. (2006). Los derechos reproductivos de las mujeres vascas en el cambio de siglo: de la anticoncepción a la reproducción asistida. *Vasconia, 35*, 345–371.

Alkorta-Idiakez, I. (2010). Three decades of reproductive rights: The highs and lows of biomedical innovations. In M. L. Esteban & M. Amurrio (Eds.), *Feminist challenges in the social sciences: Gender studies in the Basque Country* (pp. 143–158). Center for Basque Studies University of Nevada UPV/EHU.

Alkorta-Idiakez, I. (2021). Spanish legal reproscape: The making of a bio-Industry. *Law, Technology and Humans, 3*(1), 123–132. https://doi.org/10.5204/lthj.1489

Almeling, R. (2011). *Sex cells: The medical market for eggs and sperm.* University of California Press.

Alvarez, B. (2013). La maternidad: entre la decisión individual y/o la obligatoriedad social. In C. López Matheu, J. Bestard, & D. Marre (Eds.), *Maternidades, procreación y crianza en transformación* (pp. 219–242). Editorial Bellaterra.

Arregui, A. G. (2020). Positional wildness: Amazonian ribeirinhos, pink dolphins and interspecies affections. *Ethnos, 85*(5), 819–842. https://doi.org/10.1080/00141844.2019.1619606

Arribas-Ayllon, M. (2011). *Genetic testing: Accounts of autonomy, responsibility, and blame.* Routledge.

Bach, A. S., & Kroløkke, C. (2019). Hope and happy futurity in the cryotank: Biomedical imaginaries of ovarian tissue freezing. *Science as Culture, 29*(3), 425–449. https://doi.org/10.1080/09505431.2019.1681953

Baldwin, K., Culley, L., Hudson, N., & Mitchell, H. (2019). Running out of time: Exploring women's motivations for social egg freezing. *Journal of Psychosomatic Obstetrics & Gynecology, 40*(2), 166–173. https://doi.org/10.1080/0167482x.2018.1460352

Barad, K. (1998). Getting real: Technoscientific practices and the materialization of reality. *Differences: A Journal of Feminist Cultural Studies, 10*(2), 87–128.

Beeson, D., Darnovsky, M., & Lippman, A. (2015). What's in a name? Variations in terminology of third-party reproduction. *Reproductive Biomedicine Online, 31*(6), 805–814. https://doi.org/10.1016/s1472-6483(10)60647-5

Beeson, D., & Lippman, A. (2006). Egg harvesting for stem cell research: Medical risks and ethical problems. *Reproductive Biomedicine Online, 13*(4), 573–579. https://doi.org/10.1016/S1472-6483(10)60647-5

Bergmann, S. (2012). Resemblance that matters: On transnational anonymized egg donation in two european IVF clinics. In M. Knecht, M. Klotz, & S. Beck (Eds.), *Reproductive technologies as global form: Ethnographies of knowledge, practices, and transnational encounters* (pp. 280–289). Campus Verlag.

Bergmann, S. (2014). "La genética es como la masa de la pizza". El matching y la clasificación del fenotipo como práctica y decisión cultural en las clínicas de reproducción asistida. In E. Pérez Sedeño & E. Ortega Arjonilla (Eds.), *Cartografías del cuerpo: Biopolíticas de la ciencia y la tecnología* (pp. 307–352). Cátedra.

Bergmann, S. (2015). Assisted authenticity: Naturalisation, regulation and the enactment of "race" through donor matching. In V. Kantsa, G. Zanini, & L. Papadopoulou (Eds.), *(In)Fertile Citizens. Anthropological and Legal Challenges of Assisted Reproduction Technologies* (pp. 231–246). University of Aegean.

Bestard, J., & Orobitg, G. (2009). Le paradoxe du don anonyme. Signification des dons d'ovules dans les procréations médicalement assistées. In E. Porqueres i Gené (Ed.), *Défis contemporains de la parenté* (p. 277301). Éditions de L'EHESS.

Bonyad, T. (2024, 15). Framing the labor of paid egg donors in Iran: Marginality, gendered care, and divine reward. *Medical Anthropology*, 1. https://doi.org/10.1080/01459740.2024.2395291

Boris, E., & Salazar Parreñas, R. (2010). Introduction. In E. Boris & R. Salazar Parreñas (Eds.), *Intimate labors: Cultures, technologies, and the politics of care* (pp. 1–12). Stanford University Press.

Brody, J. E. (2017, July 10). Do egg donors face long-term risks? *The New York Times*. https://www.nytimes.com/2017/07/10/well/live/are-there-long-term-risks-to-egg-donors.html

Brown, W. (2015). *Undoing the demos: Neoliberalism's stealth revolution*. Zone Books.

Bühler, N., de Jong, W., Nay, Y. E., & Zehnder, K. (2015). "Ontological choreography" as an ethnographic tool: Understanding the making of families by reproductive technologies in Switzerland. *Swiss Journal of Sociocultural Anthropology, 20*, 84–96. https://doi.org/10.36950/tsantsa.2015.20.7435

Cassidy, R., & Mullin, M. (2007). *Where the wild things are now: Domestication reconsidered*. Routledge.

Cohen, L. (2005). Operability, bioavailability, and exception. In A. Ong & S. J. Collier (Eds.), *Global assemblages: Technology, politics, and ethics as anthropological problems* (pp. 79–90). Blackwell Publishing.

Cooper, M., & Waldby, C. (2014). *Clinical labor: Tissue donors and research subjects in the global bioeconomy*. Duke University Press.

Degli Esposti, S., & Pavone, V. (2019). Oocyte provision as a (quasi) social market: Insights from Spain. *Social Science & Medicine, 234*, 1–9. https://doi.org/10.1016/j.socscimed.2019.112381

Douglas, M. (1966). *Purity and danger: An analysis of concept of pollution and taboo*. Routledge.
Ducey, A. (2010). Technologies of caring labor: From objects to affect. In E. Boris & R. Salazar Parreñas (Eds.), *Intimate labors cultures, technologies, and the politics of care* (pp. 18–32). Stanford University Press.
ESHRE. (2017). *Press statements for journalists, 3. Egg donation fact sheets*. European Society of Human Reproduction and Embryology. https://www.eshre.eu/Press-Room/Resources
Fannin, M. (2020). Embodied labour in the bioeconomy. In A. Datta, P. Hopkins, L. Johnston, E. Olson, & J. Maria Silva (Eds.), *Routledge handbook of gender and feminist geographies* (pp. 326–335). Routledge.
Federici, S. (2012). *Revolution at point zero: Housework, reproduction, and feminist struggle*. PM Press.
Foucault, M. (1979). *Discipline and punish: the birth of the prison*. Peregrine.
Foucault, M. (1988). *Technologies of the self: A seminar with Michel Foucault*. University of Massachusetts Press.
Franklin, S. (1997). *Embodied progress: A cultural account of assisted conception*. Routledge.
Franklin, S. (2006). *Born and made: An ethnography of preimplantation genetic diagnosis*. Princeton University Press.
Franklin, S. (2013). *Biological relatives: IVF, stem cells, and the future of kinship*. Duke University Press.
Gentile, K. (2015). *The business of being made: The temporalities of reproductive technologies, in psychoanalysis and culture*. Routledge.
Halovic, S. (2017). The future emerges from the past: Comment on "personal genomic testing, Genetic inheritance, and uncertainty". *Journal of Bioethical Inquiry, 14*(4), 591–592. https://doi.org/10.1007/s11673-017-9818-5
Haraway, D. (2008). *When species meet*. University of Minnesota Press.
Haraway, D. J. (1991). *Simians, cyborgs and women: The reinvention of nature*. Routledge.
Himmelweit, S. (1995). The discovery of "unpaid work": The social consequences of the expansion of "work". *Feminist Economics, 1*(2), 1 19. https://doi.org/10.1080/714042229
Hochschild, A. R. (2000). Global care chains and emotional surplus value. In A. Giddens & W. Hutton (Eds.), *On the edge: Living with global capitalism* (pp. 130–146). Jonathan Cape.
Hochschild, A. R. (2003). *The managed heart: Commercialization of human feeling*. University of California Press.

Holmberg, T., Schwennesen, N., & Webster, A. (2011). Bio-objects and the bio-objectification process. *Croatian Medical Journal, 52*(6), 740–742. https://doi.org/10.3325/cmj.2011.52.740

Hovav, A. (2019). Producing moral palatability in the Mexican surrogacy market. *Gender & Society, 33*(2), 273–295. https://doi.org/10.1177/0891243218823344

Ingold, T. (2000). *The perception of the environment: Essays on livelihood, dwelling & skill.* Routledge.

Inhorn, M., Birenbaum-Carmeli, D., Birger, J., Westphal, L. M., Doyle, J., Gleicher, N., Meirow, D., Dirnfeld, M., Seidman, D., Kahane, A., & Patrizio, P. (2018). Elective egg freezing and its underlying socio-demography: A binational analysis with global implications. *Reproductive Biology and Endocrinology, 16*(1), 70. https://doi.org/10.1186/s12958-018-0389-z

Inhorn, M. C. (2015). *Cosmopolitan conceptions: IVF sojourns in global Dubai.* Duke University Press.

Jociles Rubio, M. I. (2020). "Supongo que si hubiera tenido pasta, no lo habría hecho": motivaciones para donar óvulos e ideología del altruismo. In A. M. Rivas & C. Álvarez Plaza (Eds.), *Etnografía de los mercados reproductivos: Actores, instituciones y legislaciones* (pp. 35–93). Tirant lo Blanch.

Kathryn, E., & Clare, W. (2010). A "healthy baby": The double imperative of preimplantation genetic diagnosis. *Health (London), 14*(1), 41–56. https://doi.org/10.1177/1363459309347477

Keaney, J. (2023). *Making gaybies: Queer reproduction and multiracial feeling.* Duke University Press.

Konrad, M. (2006). *Nameless relations: Anonymity, Melanesia and reproductive gift exchange between British ova donors and recipients.* Berghahn Books.

Kroløkke, C. (2014). West is best: Affective assemblages and Spanish oocytes. *European Journal Women's Studies, 21*(1), 57–71. https://doi.org/10.1177/1350506813510422

Kroløkke, C. H., & Adrian, S. W. (2013). Sperm on ice: Fatherhood and life after death. *Australian Feminist Studies, 28*(77), 263–278. https://doi.org/10.1080/08164649.2013.821728

Lacaze, P., Tiller, J., & Ryan, J. (2017). The dangers of direct-to-consumer genetic testing for Alzheimer's disease: Comment on "personal genomic testing, genetic inheritance, and uncertainty". *Journal of Bioethical Inquiry, 14*(4), 585–587. https://doi.org/10.1007/s11673-017-9817-6

Lafuente-Funes, S. (2017). *Bioeconomías reproductivas: Los óvulos en la biología pos fecundacion in vitro.* PhD Thesis, Universidad Complutense de Madrid, Madrid.

Lafuente-Funes, S. (2019). Shall we stop talking about egg donation? Transference of reproductive capacity in the Spanish Bioeconomy. *BioSocieties, 15*, 207–225. https://doi.org/10.1057/s41292-019-00149-5
Lafuente-Funes, S. (2021). *Mercados reproductivos: crisis, deseo y desigualdad.* Katakrak Liburuak.
Lafuente-Funes, S. (2023). The role of vitrification in Spanish reproductive labs: A cryo-revolution led by strategic freezing. *Science, Technology, & Human Values, 48*(4), 752–776. https://doi.org/10.1177/01622439231155649
Lafuente-Funes, S., and Pérez Orozco, A. (2020). On (global) care chains in times of crisis: Egg donation and domestic work in Spain. *: Latin American Science, Technology and Society, 3*(1), 354-376. https://doi.org/10.1080/25729861.2020.1796326
Lafuente-Funes, S., Weis, C., Hudson, N., & Provoost, V. (2023). Egg donation in the age of vitrification: A study of egg providers' perceptions and experiences in the UK, Belgium and Spain. *Sociology of Health & Illness, 45*(2), 259–278. https://doi.org/10.1111/1467-9566.13590
Latour, B. (2004). How to talk about the body? The normative dimension of science studies. *Body & Society, 10*(2–3), 205–229. https://doi.org/10.1177/1357034X04042943
Majumdar, A. (2021). Assisted reproductive technologies and the conceptualization of ageing in India. *Anthropology & Aging, 42*(1), 49–65. https://doi.org/10.5195/aa.2021.261
Mamo, L. (2005). Biomedicalizing kinship: Sperm banks and the creation of affinity-ties. *Science as Culture, 14*(3), 237–264. https://doi.org/10.1080/09505430500216833
Mansfield, B. (2017). Folded futurity: Epigenetic plasticity, temporality, and new thresholds of fetal Life. *Science as Culture, 26*(3), 355–379. https://doi.org/10.1080/09505431.2017.1294575
Marcus, G. E. (1995). Ethnography in/of the world system: The emergence of multi-sited ethnography. *Annual Review of Anthropology, 24*(1), 95–117. https://doi.org/10.1146/annurev.an.24.100195.000523
Marcus, G. E. (1998). *Ethnography through thick and thin.* Princeton University Press.
Marre, D., San Román, B., & Guerra, D. (2018). On reproductive work in Spain: Transnational adoption, egg donation, surrogacy. *Medical Anthropology, 37*(2), 158–173. https://doi.org/10.1080/01459740.2017.1361947

Martin, L. J. (2010). Anticipating infertility: Egg freezing, genetic preservation, and risk. *Gender & Society, 24*(4), 526–545. https://doi.org/10.1177/0891243210377172

Mason, P. H. (2017). Personal genomic testing, genetic inheritance, and uncertainty. *Journal of Bioethical Inquiry, 14*(4), 583–584. https://doi.org/10.1007/s11673-017-9816-7

Matorras Weinig, R., & Hernández, J. (Eds.). (2007). *Estudio y tratamiento de la pareja estéril: Recomendaciones de la Sociedad Española de Fertilidad, con la colaboración de la Asociación Española para el Estudio de la Biología de la Reproducción, de la Asociación Española de Andrología y de la Sociedad Española de Contracepción*. Adalia.

Mauss, M. (1969 [1925]). *The gift: Forms and functions of exchange in archaic societies*. Cohen and West.

Metzler, I., & Webster, A. (2011). Bio-objects and their boundaries: Governing matters at the intersection of society, politics, and science [Report]. *Croatian Medical Journal, 52*, 648–650. https://doi.org/10.3325/cmj.2011.52.648

Molas, A. (2023). Recentering labor in the egg donation bioeconomy: Egg donors' (re)productive work and subjectification in Spain. *Medical Anthropology, 43*(8), 770–783. https://doi.org/10.1080/01459740.2023.2263806

Molas, A., & Perler, L. (2020). Selecting women, taming bodies? Body ontologies in egg donation practices in Spain. *Tapuya: Latin American Science, Technology and Society, 3*(1), 396–414. https://doi.org/10.1080/25729861.2020.1781371

Molas, A., & Perler, L. (2024). Intimate voices in transnational reproductive markets. *Medical Anthropology, 43*(8), 673–681. https://doi.org/10.1080/01459740.2024.2439967

Molas, A., & Whittaker, A. (2023). Tempting luck: Temporalities and risk anticipation among egg donors in Spain. *Science, Technology, & Human Values, 49*(6), 1256–1280. https://doi.org/10.1177/01622439231164666

Moll, T. (2019). Making a match: Curating race in South African gamete donation. *Medical Anthropology, 38*(7), 588–602. https://doi.org/10.1080/01459740.2019.1643853

Moll, T. (2024). Egg providers in eGoli. *American Ethnologist, 51*(4), 568–579. https://doi.org/10.1111/amet.13350

Mutlu, B. (2024). Strategic in/visibility of Turkish egg donors: Reproductive labor, secrecy, and stigma in the transnational bioeconomy. *Medical Anthropology, 43*(8), 697–713. https://doi.org/10.1080/01459740.2024.2355959

Nahman, M. (2008). Nodes of desire. *European Journal of Women's Studies, 15*(2), 65–82. https://doi.org/10.1177/1350506807088068

Nahman, M. (2013). *Extractions: An ethnography of reproductive tourism*. Palgrave Macmillan.

Nahman, M. (2018). Migrant extractability: Centring the voices of egg providers in cross-border reproduction. *Reproductive Biomedicine & Society Online, 7*, 82–90. https://doi.org/10.1016/j.rbms.2018.10.020

Nahman, M. (2013). *Extractions: an ethnography of reproductive tourism*. Palgrave Macmillan.

Nahman, M., & Weis, C. (2023). Redefining bioavailability through migrant egg donors in Spain. *Body & Society, 29*(1), 79–109. https://doi.org/10.1177/1357034X231161319

Namberger, V. (2019). *The reproductive body at work: The South African bio-economy of egg donation*. Routledge.

Novas, C., & Rose, N. (2000). Genetic risk and the birth of the somatic individual. *Economy and society, 29*(4), 485–513. https://doi.org/10.1080/03085140050174750

Orobitg, G., Bestard, J., & Salazar, C. (2013). El cuerpo (re)productivo. Interés económico y altruismo en las experiencias de un grupo de mujeres donantes de óvulos. *Revista Andaluza de Antropología, 5*, 91–103. https://doi.org/10.12795/RAA.2013.i05.05

Pande, A. (2014). *Wombs in labor: Transnational commercial surrogacy in India*. Columbia University Press.

Pavone, V. (2017). Bio-Identification, Value Creation and the Reproductive Bioeconomy: Insights from the Reprogenetics Sector in Spain. In V. G. Pavone & Joanna (Eds.), *Bioeconomies* (pp. 129–159). Palgrave Macmillan. https://doi.org/10.1007/978-3-319-55651-2_6

Pavone, V., & Arias, F. (2011). Beyond the geneticization thesis. *Science, Technology, & Human Values, 37*(3), 235–261. https://doi.org/10.1177/0162243911411195

Pavone, V., & Lafuente Funes, S. (2018). Selecting what? Pre-implantation genetic diagnosis and screening trajectories in Spain. In A. Wahlberg & T. M. Gammeltoft (Eds.), *Selective reproduction in the 21st century* (pp. 123–148). Palgrave Macmillan.

Pearson, H. (2006). Health effects of egg donation may take decades to emerge. *Nature, 442*, 607. https://doi.org/10.1038/442607a

Perler, L., & Sánchez Pérez, T. (2024). In/Visible – A photographic journey into the lives of egg donors in Spain. *Medical Anthropology, 43*(8), 784–801. https://doi.org/10.1080/01459740.2024.2384734

Perler, L., & Schurr, C. (2020). Intimate lives in the global bioeconomy: Reproductive biographies of Mexican egg donors. *Body & Society, 27*. https://doi.org/10.1177/1357034X2093632

Pollock, A. (2003). Complicating power in high-tech reproduction: Narratives of anonymous paid egg donors. *Journal of Medical Humanities, 24*(3-4), 241–263. https://doi.org/10.1023/A:1026010504214

Rivas, A. M., Álvarez, C., & Jociles, M. I. (2018). La intervención de 'terceros' en la producción de parentesco: perspectiva de los/as donantes, las familias y la descendencia. *Un estado de la cuestión. Revista de Antropologia Social, 27*(2). https://doi.org/10.5209/raso.61850

Rivas, A. M., Lores, F., & Jociles, M. (2019). El anonimato y el altruismo en la donación de gametos: la producción de biocapital en la industria reproductiva. *Política y sociedad, 56*(3), 623–644. https://doi.org/10.5209/poso.60564

Rivas Rivas, A. M., & Jociles Rubio, M. I. (2020). ¿Ayuda o trabajo?: La perspectiva de las donantes de óvulos sobre su participación en la reproducción biológica. In A. M. Rivas & C. Álvarez Plaza (Eds.), *Etnografía de los Mercados Reproductivos: Actores, Instituciones y Legislaciones* (pp. 95–156). Tirant lo Blanch.

Rudrappa, S. (2012). India's Reproductive Assembly Line. *Contexts, 11*(2), 22–27. https://doi.org/10.1177/1536504212446456

Rudrappa, S., & Collins, C. (2015). Altruistic agencies and compassionate consumers: Moral framing of transnational surrogacy. *Gender & Society, 29*(6), 937–959. https://doi.org/10.1177/0891243215602922

Sandelowski, M., & de Lacey, S. (2002). The uses of a "disease": Interferility as rhetorical vehicle. In M. Inhorn & F. Van Balen (Eds.), *Infertility around the globe: New thinking on childlessness, gender, and reproductive technologies* (pp. 33–51). California University Press.

Schensul, S. L., Schensul, J. J., & LeCompte, M. D. (1999). *Essential ethnographic methods: Observations, interviews, and questionnaires*. AltaMira Press.

Schneider, J. (2008). Fatal colon cancer in a young egg donor: A physician mother's call for follow-up and research on the long-term risks of ovarian stimulation. *Fertility and Sterility, 90*(5), 1–5. https://doi.org/10.1016/j.fertnstert.2007.12.074

Schneider, J., Lahl, J., & Kramer, W. (2017). Long-term breast cancer risk following ovarian stimulation in young egg donors: A call for follow-up, research and informed consent. *Reproductive Biomedicine Online, 34*(5), 480–485. https://doi.org/10.1016/j.rbmo.2017.02.003

SEF. (2022). *Registro Nacional de Actividad 2022*. Sociedad Española de Fertilidad.

Siegl, V. (2018). Aligning the Affective Body. Commercial Surrogacy in Moscow and the Emotional Labour of Nastraivatsya. *Tsantsa. Journal of the Swiss Anthropological Association, 23*, 63–72. https://doi.org/10.36950/tsantsa.2018.18.7295

Smietana, M., Thompson, C., & Twine, F. W. (2018). Making and breaking families – Reading queer reproductions, stratified reproduction and reproductive justice together. *Reproductive Biomedicine & Society Online, 7*, 112–130. https://doi.org/10.1016/j.rbms.2018.11.001

Smietana, M., & Twine, F. W. (2022). Queer decisions: Racial matching among gay male intended parents. *International Journal of Comparative Sociology, 63*(5-6), 324–344. https://doi.org/10.1177/00207152221102837

Strathern, M. (1990). *The Gender of the gift: problems with women and problems with society in Melanesia*. University of California Press.

Teman, E. (2010). *Birthing a mother: The surrogate body and the pregnant self*. University of California Press.

Thompson, C. (2005). *Making parents: The ontological choreography of reproductive technologies*. MIT Press.

Thompson, C. (2009). Skin tone and the persistence of biological race in egg donation for assisted reproduction. In E. Nakano Glenn (Ed.), *Shades of difference: Why skin color matters* (pp. 131–147). Stanford University Press.

Tober, D. (2024). *Eggonomics: The global market in human eggs and the donors who supply them*. Routledge.

Tober, D., Garibaldi, C., Blair, A., & Baltzell, K. (2021). Alignment between expectations and experiences of egg donors: What does it mean to be informed? *Reproductive Biomedicine & Society Online, 12*, 1–13. https://doi.org/10.1016/j.rbms.2020.08.003

Tober, D., & Pavone, V. (2018). Las bioeconomías de la provisión de óvulos en Estados Unidos y en España: una comparación de los mercados médicos y las implicaciones en la atención a las donantes. *Revista de Antropologia Social, 27*(2), 261–286. https://doi.org/10.5209/raso.61852

Tober, D., Pavone, V., Lafuente-Funes, S., & Konvalinka, N. (2023). Eggonomics: Vitrification and bioeconomies of egg donation in the United States and Spain. *Medical Anthropology Quarterly, 37*(3), 248–263. https://doi.org/10.1111/maq.12767

Tsing, A. (2012). Unruly edges: Mushrooms as companion species: For Donna Haraway. *Environmental Humanities, 1*(1), 141–154. https://doi.org/10.1215/22011919-3610012

Tsing, A. (2015). *The mushroom at the end of the world: On the possibility of life in capitalist ruins.* Princeton University Press.

van de Wiel, L. (2020). The speculative turn in IVF: Egg freezing and the financialization of fertility. *New Genetics and Society, 39*(3), 306–326. https://doi.org/10.1080/14636778.2019.1709430

Vertommen, S., & Barbagallo, C. (2022). The in/visible wombs of the market: The dialectics of waged and unwaged reproductive labour in the global surrogacy industry. *Review of International Political Economy, 29*(6), 1945–1966. https://doi.org/10.1080/09692290.2020.1866642

Vlasenko, P. (2024). Worker-mothers between legitimation and discipline: Ambiguities in egg donation and surrogacy in Ukraine. *Medical Anthropology, 43*(8), 740–733. https://doi.org/10.1080/01459740.2024.2439965

Vora, K. (2016). *Life support: Biocapital and the new history of outsourced labor.* University of Minnesota Press.

Wade, P. (2012). Race, kinship and the ambivalence of identity. In K. Schramm, D. Skinner, & R. Rottenburg (Eds.), *Identity politics and the new genetics: Re/creating categories of difference and belonging* (pp. 79–96). Berghahn Books.

Wahlberg, A., & Gammeltoft, T. M. (2018). *Selective reproduction in the 21st century.* Springer International Publishing.

Waldby, C. (2014). "Banking time": Egg freezing and the negotiation of future fertility. *Culture, Health & Sexuality, 17*(4), 470–482. https://doi.org/10.1080/13691058.2014.951881

Waldby, C. (2019). *The oocyte economy: The changing meaning of human eggs.* Duke University Press.

Waldby, C., & Cooper, M. (2008). The biopolitics of reproduction: Post-Fordist Biotechnology and women's clinical labour. *Australian Feminist Studies, 23*(55), 57–73. https://doi.org/10.1080/08164640701816223

Waldby, C., & Mitchell, R. (2006). *Tissue economies: Blood, organs, and cell lines in late capitalism.* Duke University Press.

Weinbaum, A. E. (2022). The slave episteme in biocapitalism. *Catalyst: Feminism, theory, technoscience, 8*(1), 1–25. https://doi.org/10.28968/cftt.v8i1.35232
Weis, C. (2017). *Reproductive migrations: Surrogacy workers and stratified reproduction in St Petersburg*. PhD Thesis, De Montfort University, Leicester.
Weis, C. (2021). *Surrogacy in Russia: An ethnography of reproductive labour, stratification and migration.* .
Whittaker, A. (2018). *International surrogacy as disruptive industry in Southeast Asia*. Rutgers University Press.
Woodriff, M., Sauer, M. V., & Klitzman, R. (2014). Advocating for longitudinal follow-up of the health and welfare of egg donors. *Fertility and Sterility, 102*(3), 662–666. https://doi.org/10.1016/j.fertnstert.2014.05.037

2

Spain's Transnational Fertility Industry

The Regulation of Assisted Reproduction

Spain's first law on assisted reproduction, implemented in 1988, was one of the earliest in Europe. It came into effect under the government of the Spanish Socialist Workers' Party (PSOE), led by Felipe González, Spain's third Prime Minister after the dictatorship of Francisco Franco (1936–1975). However, assisted reproductive techniques were being used and had been normalized long before the passage of the law. The earliest sperm bank in Spain, for example, was founded 11 years earlier, in 1977, and it facilitated the birth of thousands of babies within a decade. Victoria Anna, the first "test tube baby" in Spain, was born in 1984 (four years before the law) at the Instituto Universitario Dexeus, an important private clinic in Barcelona. Four year after her birth, the first successful IVF treatment using donated eggs resulted in the birth of twins at the same clinic (Coroleu Lletget, 2011).

The 1988 law on assisted reproduction, therefore, came to mediate the expansion and direction of these techniques, arguing that they were a "matter of great responsibility" that could no longer be delegated to the

A. Molas, *Taming Egg Donors*, Health, Technology and Society,
https://doi.org/10.1007/978-981-96-5103-0_2

"free choice of scientists"[1] (Law 35/1988, p. 33373). Other major scientific achievements took place quickly after the implementation of the law, such as the clinical application of preimplantation genetic diagnosis (PGD) in 1991, the development of intra-cytoplasmic sperm injection (ICSI) in 1992, and the first births of babies from frozen eggs in 2002 (Coroleu Lletget, 2011).

Advances in assisted reproduction during the following decades made the 1988 law quickly obsolete, and it was reformed in 2003 (Law 45/2003). The renewed law aimed to expand the framework for research and to provide guidance on the problem of "supernumerary embryos."[2] However, this new law also presented important limitations for the development of assisted reproductive medicine. On the one hand, the use of supernumerary embryos for research purposes was allowed but still very restricted. On the other, it established a limit of three eggs to be extracted for every reproductive cycle, which "made difficult the ordinary practice of the techniques of assisted reproduction" (Law 14/2006, p. 2).

Consequently, three years later, the law was modified (Law 14/2006) to facilitate the use of supernumerary embryos for research, allow embryo selection through PGD to avoid genetic diseases, and anticipate the proliferation of new technologies and practices. The new law of 2006, still in effect today, did not just represent the most permissive law in Spain but also in Europe, where regulations in assisted reproduction continue to vary widely from country to country (Mínguez & García-Velasco, 2011; Präg & Mills, 2017).

Several intersecting factors make patient access easier and faster in Spain than in most neighboring European countries (Pennings et al., 2008). First, the law does not restrict access according to marital status. Single women and lesbian couples can undergo any treatment available. Second, unlike other European countries, in Spain, third-party gamete donation is anonymous, which means that, barring extraordinary circumstances, the donors are recruited and selected by the clinic with no

[1] Original text in Spanish. Fragment translated to English by the author.

[2] I am adopting here the exact terms used in the Spanish laws on assisted reproduction of 2003 and 2006 to refer to embryos that are created through IVF and cryopreserved instead of used for reproduction. This usually occurs when more embryos are produced than ultimately needed by the recipient(s) to complete their reproductive objectives.

future possibility of contact between donors and recipients (Law 14/2006, p. 8).

Finally, Spanish private clinics offer one of the highest compensation rates for egg donation across Europe (Pennings et al., 2014), an amount similar to the minimum professional gross monthly salary for a full-time job: 1080 euros in 2023. The financial compensation for third-party providers has also changed over time in Spanish regulations on assisted reproduction. The first law (1988) established that gamete donation had to be free and altruistic with no mention of economic compensation. In 1998, the annual report from the National Commission of Human Assisted Reproduction (CNRHA) argued in favor of economic compensation to avoid the establishment of a black market of eggs and to maintain the high number of donations in the country (CNRHA, 1999). The report raised the concern that compensation could become the main reason for donation, and, to prevent this outcome, the commission established an illustrative amount, subject to updates according to the rate of inflation (Zafra et al., 2012). The possibility of economic compensation is stated in the current law. The amount varies slightly from clinic to clinic, ranging from 1000 to 1300 euros.

National Register of Gamete Donors

From its first version in 1988, the Spanish law on assisted reproduction stated that the maximum permitted number of babies born from the gametes of the same donor is six. Clinics interpret this maximum as being country specific and therefore send eggs abroad when a donor has reached the maximum of six babies born in Spain from her gametes. While the number six was established to avoid problems stemming from consanguinity, it has been questioned for its lack of scientific basis (Alvarez, 2017; Janssens et al., 2015).

The 1988 law planned the creation of a national register where clinics would have to enter donors' identities and the outcomes of the donations, so that other clinics could check it when recruiting new donors. However, more than 30 years after its first formulation in 1988, professionals at fertility clinics still have no means to know how many extraction cycles a woman has undergone at other clinics or whether more than six babies have been born from her gametes. When I was doing fieldwork in 2019, some clinics were participating in the first trial of the register's preliminary version called the Information System of Human Assisted Reproduction (SIRHA), which was intended to be fully implemented by 2020. My conversations with clinicians in 2023 revealed that not all fertility clinics were participating in the registry yet. Future research on the registry's implementation is therefore needed.

Anonymity in Gamete Donation

Gamete donation in Spain is anonymous by law since 1988. The current law states that donor-conceived people and their legal representatives have the right to know general information about the donor that does not include their identity. The donor must remain anonymous unless serious medical or legal issues require the release of their identity (Law 14/2006). Europe does not have supranational regulation on the matter, and for this reason, there is a wide range of approaches across countries. Moreover, several countries have reversed the principle of anonymity over time. Sweden, for example, was the first country to lift the donor anonymity rule in 1984, followed by Austria in 1992, The Netherlands in 2000, the United Kingdom in 2005, Portugal in 2018, and, finally, France in 2022. Intermediate situations can be found, for example, in Belgium, where non-anonymous donation is possible in the case of an agreement between the parties, or in Denmark, where donors can choose whether to donate anonymously or not.

Recent reports and recommendations from bioethics committees in Spain have been pressing for the reversal of this stipulation, pointing out the interest of the children who, according to Article 8 of the European Court of Human Rights (ECHR), should have the right to know their origins as "an integral part of identity" (ECHR, 2020, p. 54). In 2016, the bioethics committee of Catalonia published a report arguing against the anonymity of gamete donation (Terribas et al., 2016). This was later followed by a Spanish bioethics committee report in January 2020, which, in the same vein, highlighted some of the current flaws of the practices in assisted reproduction (de Montalvo Jääskeläinen et al., 2020). The latter report, covered in different media outlets across Spain, provoked an immediate defensive reply from the Spanish Fertility Association in a statement that reiterated the arguments they had previously outlined in a report in support of donor anonymity (Muñoz et al., 2019). Unlike the bioethics committees mentioned, the Spanish Fertility Association is an organization of fertility experts that mainly represents the interests of the private fertility sector in Spain. With numerous members of its board of directors working as clinicians in private fertility clinic, its political and dissemination activities are aimed at safeguarding the prosperity of the industry.

The report articulates classic anxieties related to non-anonymous donations, such as the problems of filiation that identity-release donors—that is, those whose personal information is accessible to recipients—could generate, as well as the genetic essentialism that these claims supposedly reify. The report also states that lifting anonymity would lead to a reduction of the number of gamete donations, which would, in turn, cause an "economic disaster" for the private fertility sector in Spain (p. 20).

As some have already pointed out, however, clinics can no longer guarantee the anonymity of the donation to the involved parties. The rise of direct-to-consumer genetic ancestry testing and international genetic databases jeopardizes this anonymity, because children born from gamete donation and donors can be matched with biological relatives in the databases (Harper et al., 2016; Pennings, 2019).

Matching Donors and Recipients

Matching donors and recipients is mandatory according to Spanish law. Fertility clinics "must strive to ensure the closest phenotypic and immunological similarity possible between the available samples [donor gametes] and the recipient woman" (Law 14/2006, p. 6). However, even though this restricts patients from choosing donors out of a catalog, there is usually a level of negotiation with patients, who might threaten to change clinics if clinics don't comply with their requests in the matching process. As I explain in detail in Chap. 7, matching donors and recipients is perceived as challenging for clinicians, who struggle to meet patients' expectations while they try to follow "objective" criteria within a decision-making process that is inherently subjective.

Although not explicitly stated in the law, nowadays a very important part of matching between patients and donors is related to genetic carrier screening. This screening allows clinics to identify the recessive diseases that intending donors carry; this information is then used when they match donors with the male partner of the recipient couple or the sperm donor. However, this screening also identifies genetic markers for diseases that might (or might not) appear later in life. The implications of this fact will be unpacked later in the section "Rejected donors, future patients?" in Chap. 7.

At the business level, the matching process is being incorporated into new marketing strategies to recruit patients. In recent years, some big clinics have started advertising artificial intelligence tools that include facial recognition and genetic and biometric information as well as temperament screening tools to match supposedly hereditary personality traits.

Private/Public Dynamics

The reproductive industry in Spain has grown and settled within the private sector since its beginnings. The predominance of private companies in reproductive medicine is a reality both in terms of innovation and of the number of clinics and treatments conducted (De la Fuente & Requena, 2011). In 1988, when the first assisted reproduction law was established, Spain had 14 clinics that offered IVF treatments, four of which were public. This number boomed to 203 assisted reproduction centers in just 15 years, with 165 private clinics versus only 38 public centers (Alkorta-Idiakez, 2006). The latest data available reveal that in 2022 Spain had 248 IVF clinics, 203 of which were private and only 45 of which were public.

The reproductive landscape in Spain has always had major differences between the private and the public sectors in terms of treatments available, conditions of use, and waiting lists. The public health system covers IVF treatments (with women's own eggs) and artificial insemination. However, the former has a waiting list of more than a year and a half, and the latter, around six months. Moreover, unlike in private clinics where there is no official age limit for access, in the public system, women need to be under 40 years old to access these techniques.

Public health care also has a limit on the number of attempts, which varies slightly from region to region: in Catalonia, it is three for IVF and three for artificial insemination, after which patients are not eligible to try again in the public system (Departament de Salut, 2016). The public system does not offer the ROPA method (reception of oocytes from the partner), a treatment available for lesbian couples in which one provides the gametes to be fertilized in vitro with donor sperm, and the other has the embryos transferred and carries the pregnancy. Finally, and most importantly, oocyte donation is generally not offered in public centers. While the official protocol of assisted reproductive techniques in the public system lists egg donation cycles as possible, conversations with several practitioners working in both the public and private sectors confirm that the availability of oocytes is very limited. Public hospitals that offer egg donation cycles usually only permit it for women with severe health complications, such as preterm ovarian failure under 35 years old, or women who have undergone cancer treatment that affected their fertility. As Spain does not have a public egg bank, when given permission to conduct such treatments, public hospitals need to purchase eggs from private egg banks in Spain.

However, even if pregnancy via egg donation is usually achieved in private fertility clinics, the remaining procedures, which unfortunately often involve complications, take place at other health institutions. According to the 2022 report, out of the 5102 embryo transfers from fresh donated eggs that took place, 58% did not result in one or more babies born. In all, 43% did not result in a pregnancy, and of the pregnancies that occurred, 20% ended in miscarriage, abortion, or other complications that led to the termination of the pregnancy (SEF, 2022).[3]

Indeed, women undergoing IVF have significantly more chances of suffering perinatal complications such as preeclampsia, low birth weight, and preterm birth (Keukens et al., 2022; Sanders et al., 2022). The resources in place to attend these serious health complications as well as the emotional investment of women, families, and professionals

[3] The outcome of 6.5% of these pregnancies was unknown to the clinics because the recipients (many of whom come from abroad) are not obligated to keep in touch with the clinic after their treatments.

challenge the assumption that IVF happens "in the clinic." In the case of Spain, which has a comprehensive public health care system, recipients will most likely be treated in public hospitals.

Similarly, when there are side effects or complications following the extraction of eggs, egg donors are not usually treated at the clinics, which in many cases do not have the infrastructure to do so. Instead, donors are referred to the emergency rooms of public hospitals, because the donation contract does not include private health insurance.

Vitrification and Donor Egg Banking

The proliferation of freezing techniques in Spain has changed fertility clinics' practice significantly in recent years (Hudson et al., 2020; Quaas & Pennings, 2018; Tober et al., 2023; Tober & Pavone, 2018; Waldby, 2019). While sperm has been frozen, stored, and shipped for decades in Europe, the much more complex technique of oocyte freezing (technically called "vitrification") has taken longer to be controlled and routinized. This routinization is now visible in the latest Spanish fertility report, which shows that 41% of egg donation cycles performed in Spain in 2022 used vitrified donated eggs (SEF, 2022).

As I observed during my fieldwork, all the clinics that had an egg donation program in place had created their own egg bank in recent years. While some used it solely to meet their own treatment demands, others also used it for shipping ova to neighboring clinics that didn't recruit egg donors. It was also common for big clinics to export vitrified donated eggs abroad. This is the case of Ovobank, a branch of a fertility clinic in Marbella that, after years of growth and expansion to other cities in Spain, established in 2013 the first commercial egg bank in Spain. Eggs are exported within and outside Europe with a special focus on Italy and Brazil, which have become important clients for Spanish egg banks (Machin et al., 2024; Tober & Pavone, 2018). According to my interviews with clinicians, in 2019 the average price per egg is 450 euros, and they are usually distributed in batches of different amounts depending on the preference of the patients or clinics abroad.

Egg vitrification techniques have many advantages for clinics. On the one hand, they avoid the need to synchronize donors and recipients in space and time, which facilitates both egg donor and patient recruitment and treatment. And on the other, they allow clinics to divide the eggs of a single donation cycle between more than one recipient, reducing waste and significantly increasing the profits that a single egg donation cycle can generate. Finally, they have the potential to reduce waiting lists even more, especially for recipients with "rare" phenotypes who might experience difficulties to find a matching donor at the same time they want to undergo the treatment (Hudson et al., 2020; Lafuente-Funes, 2023).

Egg banking, however, also raises important questions at the level of local regulations for ARTs, which egg shipping can potentially circumvent. Economic compensation for egg donors or the anonymity of donation—both banned in various European countries—could find a way to enter the legal landscape of other jurisdictions if specific Europe-level regulations on the matter are not developed.[4]

Egg shipping can ultimately change cross-border reproductive mobilities in Europe, with the potential of creating a reverse circuit in which the eggs—and not women/couples—travel. At the time of writing, the number of eggs exported abroad from Spain is not accounted for in the Sociedad Española de Fertilidad registers, meaning that new reports on cross-border reproductive care could give the misleading impression that the pressure on egg donors has been alleviated in Spain and distributed across Europe. These shipments should be counted, because the practice is likely to increase in the coming years.

European Fertility Struggles and the Egg Donation Industry

The expansion and normalization of reproductive technologies across Europe has an inverse relationship to the decline of total fertility rates across the continent (Illustrations 2.1 and 2.2). Although data on fertility rates and especially on IVF cycles is difficult to trace due to the lack of

[4] The EU Tissue and Cells Directive (2004/23/ED) regulates the obtention and circulation of cells and tissue within the European Union and states that egg procurement is only possible for therapeutic reasons, without the possibility of commercialisation or incentives for donation. As observed in the case of egg donation, however, the application and interpretation of this directive across the EU varies widely from country to country.

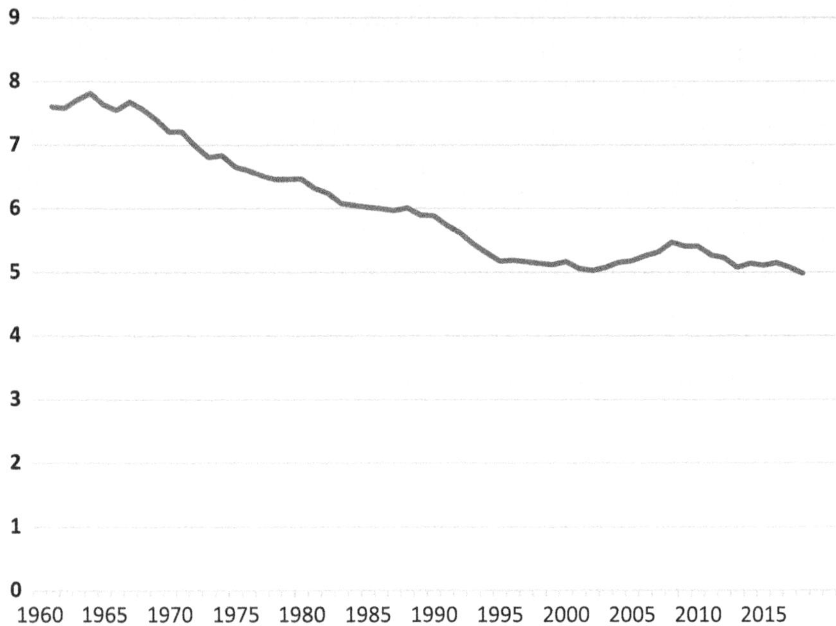

Illustration 2.1 Number of live births, EU-28, 1961–2018

early records in some European countries (Präg & Mills, 2017), recent compilation studies give us approximate numbers. While in 1980, there were 6.47 million babies born live in Europe (EU-28), this number dropped to 4.97 in 2018 (Eurostat, 2021). Contrary to this trend, in 1997 there were 203,225 treatment cycles reported in Europe, which more than quadrupled in 19 years, reaching 1,007,598 by 2018 (Wyns et al., 2022).[5]

The availability of contraceptive methods in most European countries since the 1960s and the full introduction of women into the paid labor market (te Velde et al., 2012) are usually emphasized to explain the sharp declines of fertility rates across the continent, a situation leading to what

[5] Notably, from 1997 not only the number of cycles increased, but also the number of countries reporting data to the European Society of Human Reproduction and Embryology (ESHRE). The most accurate part of the graph is probably from 2013 to 2018, where the number of countries reporting results is the most stable (see Wyns et al., 2022).

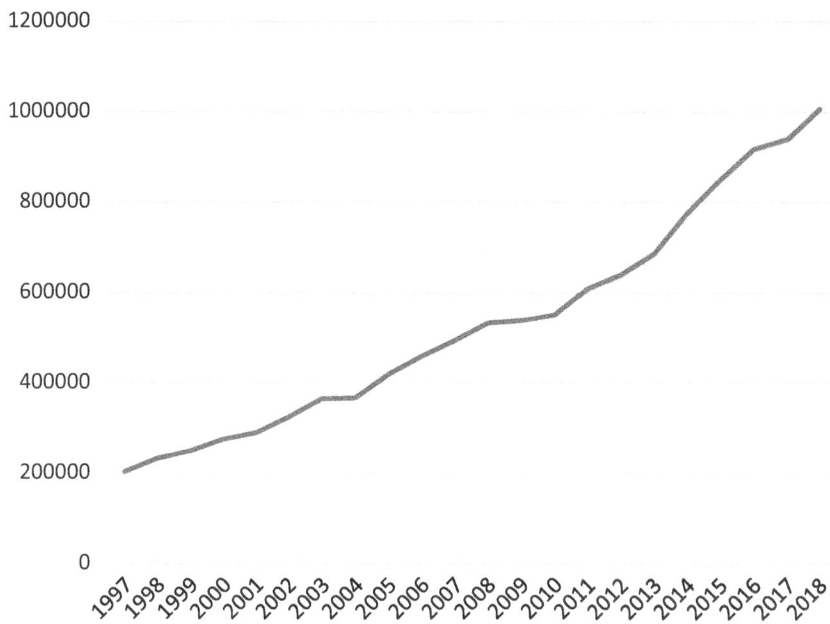

Illustration 2.2 Number of cycles in Europe, 1997–2018

Diana Marre has labeled "structural infertility" (2009, p. 114). Low fertility rates and the delay of motherhood are directly linked to difficulties in balancing work and parenthood, due to precarious and inflexible working conditions, limited access to childcare services, the still-large gap between the amount of unpaid work done by men and women, as well as cultural factors related to the social acceptance of childlessness (Alkorta-Idiakez, 2006, 2010; Inhorn et al., 2018; Matysiak et al., 2018; Sobotka, 2017; Tobío Soler, 2002; Waldby, 2019). The consequences of these changes can be seen in some representative trends, such as the rise in first-time mothers over 40 years old both in the US and Europe (Sobotka & Beaujouan, 2018). Supporting the correlation between the delay of motherhood and the demand for IVF, Spanish records indicate that 75% of the cycles using fresh donated egg in Spanish clinics are directed to women over 40 years old (SEF, 2022) (Illustration 2.3).

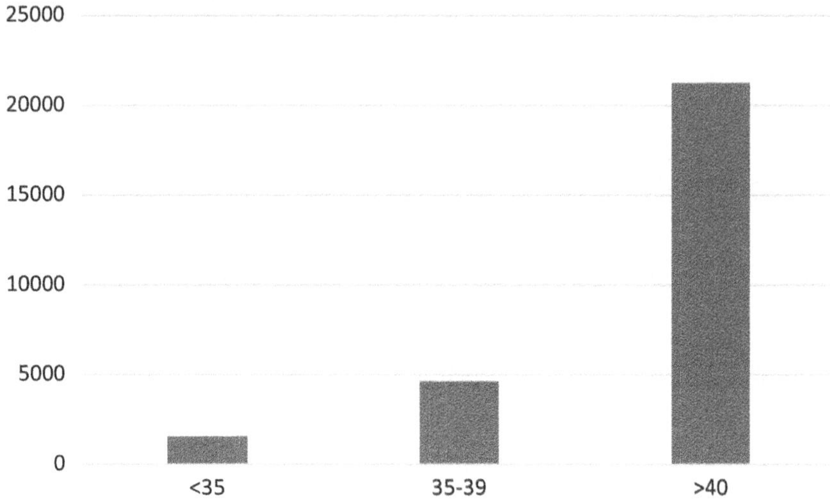

Illustration 2.3 Age of women receiving embryo transfers using egg donation in Spain

The delay in motherhood due to these intersecting factors has led to a greater number of couples with difficulties conceiving around Europe, who in turn face barriers in most of their home countries to access the desired assisted reproduction treatment, particularly IVF using donated eggs. This has generated what has been called "cross-border reproductive care" or "fertility tourism" (Präg & Mills, 2017; Sobotka, 2017; Sobotka et al., 2008; Sobotka & Beaujouan, 2018).

The latest data available at the time of writing indicate that in 2022, 27,564 patients from other countries came to Spain to undergo assisted reproduction treatments, of whom 52.3% came to undergo IVF with donated eggs. That year, Spain performed 10,849 egg extractions on women's bodies, which were used for 12,110 IVF treatments within its 242 fertility clinics. On top of that, 34,296 IVF egg donation cycles were performed using vitrification techniques, adding up to a total of 46,406 cycles using donated eggs.

"Cross-border reproductive care" vs. "fertility tourism"

Debates abound about the proper term for people who cross borders to gain access to assisted reproductive treatments. In their review of cross-border reproductive care studies, Hudson et al. (2011) point out that since the phenomenon of people crossing borders for fertility treatments was understood "as a variant of the wider trend of medical tourism" (2011, p. 678), the term "fertility tourism" was initially unproblematic. This term is still largely used today both in the media and academia, although in recent academic literature, it usually appears along with "cross-border reproductive care." Nevertheless, some scholars have started to question the word "tourism" because it implies "choice, pleasure and relaxation" (2011, p. 678). In this line, Marcia Inhorn and Pascale Patrizio (2009) point out that the tourism connotation fails to recognize the suffering and stress of those undergoing fertility treatments. Instead, the term "reproductive exile," which highlights the failure of states to meet infertile citizens' needs, has been proposed by some (Inhorn & Patrizio, 2009; Matorras, 2005). Following Guido Pennings' (2005) argument, Marcia Inhorn and Pascale Patrizio (2009) among others, advocate for more neutral terms such as "reproductive travel" or "cross-border reproductive care." The latter is the one used by the European Society of Human Reproduction and Embryology (ESHRE) (Pennings et al., 2008).

A question of sensitivity toward reproductive travelers is at the core of the debate. Indeed, as has been repeatedly demonstrated, fertility patients suffer, and most of them travel against their wishes, for reasons related to legal restrictions and availability of treatments that have little to do with the consumerist values that the concept of "tourism" entails (Pennings et al., 2008). The experiences of couples traveling to Spain to "circumvent laws" (Bergmann, 2011) have been the focus of different ethnographic studies (Bergmann, 2011, 2012; Kroløkke, 2014a, 2014b; Zanini, 2011, 2013).

However, I contend that we could expand the discussion by asking if the term used to define reproductive travelers can or should depend just on patients' intentions, motivations, and feelings. I argue that different types of reproductive circuits might need different names to describe their nature and particularities and that these names should consider not just the patients' perspective, but also the logics embedded in these mobilities and their impact in the recipient country. I cast doubt as well on the claim that we have to aim for an "objective" term with "no value judgment regarding the movements" (Pennings, 2005, p. 3571), as I believe the topic touches sensitive issues that need to be addressed critically by social researchers. But before moving on, I consider the data that shapes my argument.

Spain at the Core of Mobilities for Cross-Border Reproductive Care

An analysis of the records of 40 European countries in 2019 reported by the ESHRE (Smeenk et al., 2023) show that in Europe, the "objective" (Pennings, 2005, p. 3571) term "cross-border reproductive care" describes what is in practice a much more complex phenomenon. Although the authors acknowledge that the data provided by these countries have varying levels of completeness, the results still point to interesting trends. Cross-referencing the data from this report with that of the report of the activity of Spanish fertility clinics in 2019 (SEF, 2019), we find that 18,475 of the 33,003 cycles for cross-border patients reported in 13 European countries were conducted in Spain, which represents 56% of all the reproductive mobility in Europe. We must recognize, therefore, that when discussing cross-border reproductive care in Europe, we are mainly talking about infertile European patients traveling to Spain.

Eggs at the Core of Cross-Border Reproductive Care Mobilities Toward Spain

Departing from the same ESHRE report (Smeenk et al., 2023) and matching its data against the report of activity of Spanish fertility clinics in 2019 (SEF, 2019), a second interesting aspect emerges. Fifty-two point nine percent of these European mobilities occur to gain access to egg donation treatments. If we consider this number in the light of the 10,024 foreign couples that underwent IVF with donated eggs in Spain in 2019, we can conclude that 57% of cross-border egg donation treatments in Europe happened in Spain. A second important consideration, therefore, is that the main motivation to travel abroad for fertility treatment is the availability of ova, which in practice means that most European international fertility patients travel to Spain for donated eggs.

Economic Precariousness and Egg Availability in Spain

Although Spain is the hub of ova extraction in Europe, very little quantitative research has been done to understand the sociodemographic characteristics of women who donate their eggs. The biggest survey done to date was undertaken by the ESHRE with data from 2011 and 2012 from a total of 449 Spanish egg donors (Pennings et al., 2014). The study shows that 28% of participants worked full-time, while 23.3% worked part-time, 23.3% were unemployed, and 24.9% were studying. Forty-six point seven percent of donors had completed secondary studies as their highest level of education, while 22.5% had completed technical studies and 22.5% university studies. The study also asked former donors about their motivations for donating and pointed out that 56.5% claimed to be motivated by both altruism and financial reasons, while 30.4% claimed being motivated only by altruism and 19.1% only by financial reasons. There was a strong correlation between donors' country of origin, age, professional activity, and number of donation cycles with the motivations for donating. Donors who were migrants, young, lacking a full-time job, and/or who donated more than once were more likely to be totally or partially economically motivated.

The Catalan health authorities also formerly gathered details on egg donors' sociodemographic profiles along with information about the activity of clinics in Catalonia. However, in 2016 they stopped doing so after the Spanish Health Ministry announced the creation of SIRHA, a more complete mandatory registry which would include a donors' registry, a biovigilance registry, and a description of the clinics' activity at the national level. To date, however, no sociodemographic data on donors can be found in the yearly published reports. The last Catalan report, with data from 2014, states that most egg donors were between 18 and 24 years old (40.6%). It also shows that 24.8% of egg donors were born abroad, which suggests that the oocyte economy in Spain is not just fueled by foreign couples circumventing laws, but also by migrant women who participate in large numbers in this industry (FIVCAT.NET, 2014). These data, however, are much better understood when analyzed alongside the socioeconomic context of young women in Spain. In the following section, I provide an overview of this context.

The Spanish Socioeconomic Context

The global financial crisis in 2008 hit Spain especially hard with unemployment rates reaching 25.7% in 2012 (INE, 2023a). This downfall affected the country as a whole but had a special impact among women, young adults, and older adults. The austerity measures demanded by the European Union for the management of the crisis resulted in increased precariousness of the labor market as well as major cuts in public services such as health, education, and pensions.

Although surveys show how, within the last decade, the median salary has been increasing slightly, from 1774 euros gross per month in 2008 to 2087 euros gross per month in 2021 (with minimum gross salaries at 700 euros per month in 2008 and 1080 in 2023), this has not been the case for young people. The median salary for women aged 16–24 has been stable over the last decade and has even slightly decreased in some periods such as in 2016 at 1029 euros gross per month (INE, 2023b). Moreover, the data reflect only young women who are active in the paid labor market, but 46.05% of women aged 16–19 are unemployed and 28.42% of women aged 20–24 are unemployed (INE, 2023a).

Importantly, while the minimum salary in Spain in 2023 was 15,120 euros gross per year, which represented a significant increase in relation to 2018, at 10,302.6 euros gross per year (Eurostat, 2023), these amounts reflect the full-time regular contracts of 40 hours per week. They do not include the types of contracts usually held by young people, such as training contracts, in which the salary has to be at least 75% of the official minimum wage during the first year, and 85% from the second year onward. Nor do they account for internship contracts, which are usually unpaid. They also don't account for short-term, part-time, and self-employment contracts. It is within this context of flexible, unstable, and poorly paid employment for young women that the amount of money offered in economic compensation for egg donation in Spain becomes attractive.

The Landscape of Spanish Extractive Tourism and Fertility

Given that most fertility mobilities across Europe are from patients traveling to Spain because of the availability of ova and that egg donors in Spain are mainly young women (both local and migrant) in precarious contexts at least partially driven by the economic compensation, I argue that regardless of whether we call these mobilities "cross-border reproductive care" or "fertility tourism," the fertility and tourism industries in Spain are intimately related. As I unpack further below, fertility and tourism share an unequal distribution across Europe, a transnational business structure, and the participation of poorly paid workers, often migrants and women, who provide this labor.

Although Spain is the country with the most prominent fertility industry in Europe, the industry's infrastructure is not spread equally across the territory. One hundred fifty-eight out of the 242 fertility clinics in Spain are situated along the Mediterranean coast (Andalucía, Catalonia, Valencia, and Murcia), or in Madrid, the capital. These locations are among the parts of Spain most visited by tourists (INE, 2023c) and have considerable tourism infrastructures, such as airports with low-cost flights connected to most European capitals, numerous accommodation options, as well as services in multiple languages, especially in the big cities where most clinics are based. This is not so much to suggest that patients choose these destinations for their tourist interest, but rather that this tourism infrastructure is crucial for the European transnational fertility industry to work efficiently.

Fertility Clinics as Transnational Corporations

We must also consider the fact that the Spanish reproductive industry is not a nationally based economy. Rather, it is increasingly part of the multibillion-dollar transnational private health care industry. The laissez-faire politics of the Spanish government in relation to the fertility industry has acted as a "guarantor of the profits of the neoliberal system" (Lafuente-Funes, 2017, p. 83) and has driven the Spanish fertility industry, and particularly the Spanish oocyte economy, to become an important part of the prominent reproductive bioeconomy.

Two of the biggest IVF groups in Spain—Eugin and IVI—provide good examples of the ongoing functioning of this prominent bioeconomy, which is increasingly organized transnationally. Eugin is a fertility clinic group founded in 1999 in Barcelona that currently performs the most cycles across Spain. After expanding its clinics across Europe and Latin America, it was acquired in 2015 by the biggest private health care group in the United Arab Emirates, NMC Health, which is listed on the London Stock Exchange (Casado, 2015). NMC Health later sold Eugin to the German group Fresenius Helios in 2020, owner of the biggest private health care groups in Spain (Muñoz, 2020). Another example is IVI, a private fertility group that has 30 clinics in the main cities of Spain. In 2017, IVI merged with the US private reproductive health group, RMA, becoming the biggest group offering fertility services in the world (Zafra, 2017). In 2023, IVI was bought again by the American investment company KKR for 3 billion euros (Bayón, 2023a). Eugin and IVI two fertility groups finally merged together after a long negotiation that concluded with IVI buying the American and Canadian clinics of Eugin from Fresenius Helios and the investment fund GED Capital buying the Eugin clinics in Spain and Latin America in a transaction costing 500 million euros (Bayón, 2023b).

Within this globalized bioeconomy, Lorraine Culley and Nicky Hudson (2010) argue that fertility travelers can be theorized as "global consumers, engaged in a worldwide phenomenon which is characterized by the extension of markets into human reproduction; the growing commodification of the body and its parts, and the actions of the state and other political actors in (often indirectly and perversely) aiding the development of a global market in fertility treatment" (p. 142). As they state, cross-border reproductive care is "an expression of the market par excellence" (p. 146; see also Whittaker & Speier, 2010). In Spain, the fertility industry operates in a complex dynamic embedded in a neoliberal free-market system, which works within the logics of globalization and is organized transnationally rather than locally, through intimate relations with other prominent industries. Within this framework, patient mobility and the Spanish tourism infrastructure are intrinsically related and fuel each other. Therefore, I contend that reintroducing the term "tourism" into the definition of fertility travel in Europe could have an important function even if it does not reflect the patients' experiences. Instead, it works to highlight the intimate connections between fertility mobility and the tourism circuits of Europe—and Spain's dominance in both. It also provocatively associates the transnational fertility industry with the struggles of tourist cities in Spain, such as Barcelona, where the dynamics of extraction are contributing to increase the already significant precariousness of citizens.

Conclusions

In this chapter, I have outlined contextual data that is key to understanding the intersecting factors that made Spain the main provider of ova and egg donation treatments around Europe. First, I have provided details on the past and current legislation and practices around ARTs in Spain, which have led to the most permissive regulations across Europe. Second, I have focused on the low fertility rates in Europe as well as the mobility patterns of patients across borders to access IVF. I have done so to shed light on the debates around the terminology used to describe these movements. By providing key information about the dynamics of the fertility industry and the tourism industry in Spain, I have shown how the two are similar and, in practice, intimately linked.

In the following chapter, I focus on women's decision-making to become egg donors and the close relationship between this process and the unstable and flexible labor market in Spain. I detail how the dichotomy between altruism and economic interest is revealed not only in women's narratives about donation, but also in their ambivalence and shame for not aligning (or not entirely aligning) with the gendered principle of care and solidarity in which egg donation is framed in Spain.

References

Alkorta-Idiakez, I. (2006). Los derechos reproductivos de las mujeres vascas en el cambio de siglo: de la anticoncepción a la reproducción asistida. *Vasconia, 35*, 345–371.

Alkorta-Idiakez, I. (2010). Three decades of reproductive rights: The highs and lows of biomedical innovations. In M. L. Esteban & M. Amurrio (Eds.), *Feminist challenges in the social sciences: Gender studies in the Basque Country* (pp. 143–158). Center for Basque Studies University of Nevada UPV/EHU.

Alvarez, J. G. (2017). Artículo 5. Donantes y contratos de donación. In J. A. Cobacho Gómez & J. J. Iniesta Delgado (Eds.), *Comentarios a la Ley 14/2006, de 26 de mayo, sobre Técnicas de Reproducción Humana Asistida*. Thomson-Aranzadi.

Bayón, Á. (2023a, January 13). El fondo KKR firma la compra de IVI por 3.000 millones tras once meses de espera. *Cinco Días*. https://cincodias.elpais.com/cincodias/2023/01/11/companias/1673465860_730290.html

Bayón, Á. (2023b, November 8). KKR y GED compran a Fresenius las clínicas de reproducción asistida Eugin. *Cinco Días*. https://cincodias.elpais.com/companias/2023-11-08/kkr-y-ged-compran-a-fresenius-las-clinicas-de-reproduccion-asistida-eugin.html

Bergmann, S. (2011). Reproductive agency and projects: Germans searching for egg donation in Spain and the Czech Republic. *Reproductive Biomedicine Online, 23*(5), 600–608. https://doi.org/10.1016/j.rbmo.2011.06.014

Bergmann, S. (2012). Cambiar el óvulo manteniendo el fenotipo: la búsqueda de semejanza en la FIV con donación de óvulos transnacional. In E. Pérez & R. Ibáñez (Eds.), *Cuerpos y diferencias*. Plaza y Valdés.

Casado, R. (2015, February 24). NMC Health compra la Clínica Eugin de Barcelona a ProA Capital por 143 millones. *Expansion*. https://www.expansion.com/2015/02/24/empresas/1424766054.html

CNRHA. (1999). *Primer informe anual de 1998 (Conclusiones)*. Comisión Nacional de Reproducción Humana Asistida, Ministerio de Sanidad, Gobierno de España. https://archivos.juridicas.unam.mx/www/bjv/libros/5/2291/3.pdf

Coroleu Lletget, B. (2011). Orígenes, Antecedentes e Hitos más Importantes de la Especialidad en España. In R. Matorras Weinig (Ed.), *Libro blanco sociosanitario "La infertilidad en España: Situación Actual y Perspectivas"* (pp. 71–83). Sociedad Española de Fertilidad.

Culley, L., & Hudson, N. (2010). Fertility tourists or global consumers? A sociological agenda for exploring cross-border reproductive travel. *International Journal of Interdisciplinary Social Sciences, 4*(10), 139–150. https://doi.org/10.18848/1833-1882/CGP/v04i10/53016

De la Fuente, P. E., & Requena, A. (2011). Situación y Tendencias en la Sanidad Privada. In R. Matorras Weinig (Ed.), *Libro blanco sociosanitario "La infertilidad en España: Situación Actual y Perspectivas"* (pp. 163–174). Sociedad Española de Fertilidad.

de Montalvo Jääskeläinen, F., Altisent Trota, R., Bellver Capella, V., Cadena Serrano, F., de los Reyes López, M., de la Gándara del Castillo, Á., Guillén Navarro, E., Jouve de la Barreda, N. S., López Moratalla, N., Ruiz Sicilia, L., Serrano Ruiz-Calderón, J. M., & Sánchez Chamorro, E. (2020). *Informe del Comité de Bioética de España sobre el derecho de los hijos nacidos de las técnicas de reproducción humana asistida a conocer sus orígenes biológicos*. Comité de Bioética de España.

Departament de Salut. (2016). *Protocol de les tècniques de reproducció humana assistida*. Departament de Salut, Generalitat de Catalunya.

ECHR. (2020). *Guide on article 8 of the European Convention on Human Rights*. European Convention on Human Rights, Council of Europe.

Eurostat. (2021). *Live births (total) by month*. https://ec.europa.eu/eurostat/databrowser/view/demo_fmonth/default/table?lang=en

Eurostat. (2023). Minimum wages. https://ec.europa.eu/eurostat/statistics-explained/index.php?title=Minimum_wage_statistics

FIVCAT.NET. (2014). *Estadística de la reproducció humana assistida a Catalunya, 2014*. Generalitat de Catalunya. Departament de Salut. https://scientiasalut.gencat.cat/handle/11351/3115?locale-attribute=es

Harper, J. C., Kennett, D., & Reisel, D. (2016). The end of donor anonymity: How genetic testing is likely to drive anonymous gamete donation out of business. *Human Reproduction, 31*(6), 1135–1140. https://doi.org/10.1093/humrep/dew065

Hudson, N., Culley, L., Blyth, E., Norton, W., Rapport, F., & Pacey, A. (2011). Cross-border reproductive care: A review of the literature. *Reproductive Biomedicine Online, 22*(7), 673–685. https://doi.org/10.1016/j.rbmo.2011.03.010

Hudson, N., Culley, L., Herbrand, C., Pavone, V., Pennings, G., Provoost, V., Coveney, C., & Funes, S. L. (2020). Reframing egg donation in Europe: New regulatory challenges for a shifting landscape. *Health Policy and Technology, 9*(3), 308–313. https://doi.org/10.1016/j.hlpt.2020.04.007

INE. (2023a). *Encuesta de Población Activa. Segundo trimestre 2019*. https://www.ine.es/jaxiT3/Datos.htm?t=10896

INE. (2023b). *Ganancia media anual por trabajador. Sexo y edad*. https://www.ine.es/jaxiT3/Datos.htm?t=10896

INE. (2023c). *Movimiento Turísticos en Fronteras. Número de turistas según comunidad autónoma de destino principal*. https://www.ine.es/jaxiT3/Datos.htm?t=23988#!tabs-tabla

Inhorn, M., Birenbaum-Carmeli, D., Birger, J., Westphal, L. M., Doyle, J., Gleicher, N., Meirow, D., Dirnfeld, M., Seidman, D., Kahane, A., & Patrizio, P. (2018). Elective egg freezing and its underlying socio-demography: A binational analysis with global implications. *Reprod Biol Endocrinol, 16*(1), 1–11. https://doi.org/10.1186/s12958-018-0389-z

Inhorn, M. C., & Patrizio, P. (2009). Rethinking reproductive "tourism" as reproductive "exile". *Fertility and Sterility, 92*(3), 904–906. https://doi.org/10.1016/j.fertnstert.2009.01.055

Janssens, P. M., Thorn, P., Castilla, J. A., Frith, L., Crawshaw, M., Mochtar, M., Bjorndahl, L., Kvist, U., & Kirkman-Brown, J. C. (2015). Evolving minimum standards in responsible international sperm donor offspring quota. *Reproductive Biomedicine Online, 30*(6), 568–580. https://doi.org/10.1016/j.rbmo.2015.01.018

Keukens, A., van Wely, M., van der Meulen, C., & Mochtar, M. H. (2022). Pre-eclampsia in pregnancies resulting from oocyte donation, natural conception or IVF: A systematic review and meta-analysis. *Human Reproduction, 37*(3), 586–599. https://doi.org/10.1093/humrep/deab267

Kroløkke, C. (2014a). Eggs and euros: A feminist perspective on reproductive travel from Denmark to Spain. *International Journal of Feminist Approaches to Bioethics, 7*(2), 144–163. https://doi.org/10.3138/ijfab.7.2.144

Kroløkke, C. (2014b). West is best: Affective assemblages and Spanish oocytes. *European Journal Women's Studies, 21*(1), 57–71. https://doi.org/10.1177/1350506813510422

Lafuente-Funes, S. (2017). *Bioeconomías reproductivas: Los óvulos en la biología pos fecundacion in vitro.* PhD Thesis, Universidad Complutense de Madrid, Madrid.

Lafuente-Funes, S. (2023). The role of vitrification in Spanish reproductive labs: A cryo-revolution led by strategic freezing. *Science, Technology, & Human Values*. https://doi.org/10.1177/01622439231155649

Machin, R., Álvarez Plaza, C., & Puig Hernández, M. A. (2024). The reproductive silk route: Transnational mobility of oocytes from Europe to Brazil. *Mobilities, 19*(2), 282–295. https://doi.org/10.1080/17450101.2023.2220976

Marre, D. (2009). Los silencios de la adopción en España. *Revista de Antropologia Social, 18*, 97–126.

Matorras, R. (2005). Reproductive exile versus reproductive tourism. *Human Reproduction, 20*(12), 3571–3571. https://doi.org/10.1093/humrep/dei223

Matysiak, A., Sobotka, T., & Vignoli, D. (2018). *The Great Recession and fertility in Europe: A sub-national analysis.* Vienna Institute of Demography Working Papers.

Mínguez, Y., & García-Velasco, J. A. (2011). Situación Jurídica: Comparación con Países del Entorno. In R. Matorras Weinig (Ed.), *Libro blanco sociosanitario "La infertilidad en España: Situación Actual y Perspectivas"* (pp. 97–109). Sociedad Española de Fertilidad.

Muñoz, A. (2020, December 10). Fresenius, dueño de Quirónsalud, derrota a CVC y ultima la compra de las clínicas Eugin por 350 millones. *elEconomista.es.* https://www.eleconomista.es/empresas-finanzas/noticias/10936341/12/20/Fresenius-dueno-de-Quironsalud-derrota-a-CVC-y-ultima-la-compra-de-las-clinicas-Eugin-por-350-millones.html

Muñoz, M., Abellán-García, F., Cuevas, I., de la Fuente, A., Iborra, D., Mataró, D., Núñez, R., & Roca, M. (2019). *Documento Sobre Posicionamiento de la Sociedad Española de Fertilidad Respecto de la Regla del Anonimato en las Donaciones de Gametos.* Sociedad Española de Fertilidad.

Pennings, G. (2005). Reply: Reproductive exile versus reproductive tourism. *Human Reproduction, 20*(12), 3571–3572. https://doi.org/10.1093/humrep/dei224

Pennings, G. (2019). Genetic databases and the future of donor anonymity. *Human Reproduction, 34*(5), 786. https://doi.org/10.1093/humrep/dez029

Pennings, G., de Mouzon, J., Shenfield, F., Ferraretti, A. P., Mardesic, T., Ruiz, A., & Goossens, V. (2014). Socio-demographic and fertility-related characteristics and motivations of oocyte donors in eleven European countries. *Human Reproduction, 29*(5), 1076–1089. https://doi.org/10.1093/humrep/deu048

Pennings, G., de Wert, G., Shenfield, F., Cohen, J., Tarlatzis, B., & Devroey, P. (2008). ESHRE Task Force on Ethics and Law 15: Cross-border reproductive care. *Human Reproduction, 23*(10), 2182–2184. https://doi.org/10.1093/humrep/den184

Präg, P., & Mills, M. C. (2017). Assisted reproductive technologies in Europe: Usage and regulation in the context of cross-border reproductive care. In M. Kreyenfeld & D. Konietzka (Eds.), *Childlessness in Europe: Contexts, causes, and consequences.* Springer. https://doi.org/10.1007/978-3-319-44667-7_14

Quaas, A. M., & Pennings, G. (2018). The current status of oocyte banks: Domestic and international perspectives. *Fertility and Sterility, 110*(7), 1203–1208. https://doi.org/10.1016/j.fertnstert.2018.07.013

Sanders, J. N., Simonsen, S. E., Porucznik, C. A., Hammoud, A. O., Smith, K. R., & Stanford, J. B. (2022). Fertility treatments and the risk of preterm birth among women with subfertility: A linked-data retrospective cohort study. *Reproductive Health, 19*(1), 83. https://doi.org/10.1186/s12978-022-01363-4

SEF. (2019). *Registro Nacional de Actividad 2019.* Sociedad Española de Fertilidad.

SEF. (2022). *Registro Nacional de Actividad 2022.* Sociedad Española de Fertilidad.

Smeenk, J., Wyns, C., De Geyter, C., Kupka, M., Bergh, C., Cuevas Saiz, I., De Neubourg, D., Rezabek, K., Tandler-Schneider, A., Rugescu, I., & Goossens, V. (2023). ART in Europe, 2019: Results generated from European registries by ESHRE. *Human Reproduction, 38*(12), 2321–2338. https://doi.org/10.1093/humrep/dead197

Sobotka, T. (2017). Childlessness in Europe: Reconstructing long-term trends among women born in 1900–1972. In M. Kreyenfeld & D. Konietzka (Eds.), *Childlessness in Europe: Contexts, causes, and consequences* (pp. 17–53). Springer. https://doi.org/10.1007/978-3-319-44667-7_2

Sobotka, T., & Beaujouan, É. (2018). Late motherhood in low-fertility countries: Reproductive intentions, trends and consequences. In D. Stoop (Ed.), *Preventing age related fertility loss* (pp. 11–29). https://doi.org/10.1007/978-3-319-14857-1_2

Sobotka, T., Hansen, M. A., Jensen, T. K., Pedersen, A. T., Lutz, W., & Skakkebæk, N. E. (2008). The contribution of assisted reproduction to completed fertility: An analysis of Danish data. *Population and Development Review, 34*(1), 79–101. https://doi.org/10.1111/j.1728-4457.2008.00206.x

te Velde, E., Habbema, D., Leridon, H., & Eijkemans, M. (2012). The effect of postponement of first motherhood on permanent involuntary childlessness and total fertility rate in six European countries since the 1970s. *Human Reproduction, 27*(4), 1179–1183. https://doi.org/10.1093/humrep/der455

Terribas, N., Farnós, E., Boada, M., Broggi, M. A., Busquets Font, J. M., Calaf, J., Guerra, D., Hernández, E., María, F. J., Mayoral, J., Pardo, M., & Román, B. (2016). *Dret a conèixer orígens biològics i genètics de la persona*. Comitè de Bioètica de Catalunya.

Tober, D., & Pavone, V. (2018). Las bioeconomías de la provisión de óvulos en Estados Unidos y en España: una comparación de los mercados médicos y las implicaciones en la atención a las donantes. *Revista de Antropologia Social, 27*(2), 261–286. https://doi.org/10.5209/raso.61852

Tober, D., Pavone, V., Lafuente-Funes, S., & Konvalinka, N. (2023). Eggonomics: Vitrification and bioeconomies of egg donation in the United States and Spain. *Medical Anthropology Quarterly, 37*(3), 248–263. https://doi.org/10.1111/maq.12767

Tobío Soler, C. (2002). Conciliación o contradicción: cómo hacen las madres trabajadoras. *Reis: Revista española de investigaciones sociológicas, 97*(2), 155–186.

Waldby, C. (2019). *The oocyte economy: The changing meaning of human eggs*. Duke University Press.

Whittaker, A., & Speier, A. (2010). "Cycling overseas": Care, commodification, and stratification in cross-border reproductive travel. *Medical Anthropology, 29*(4), 363–383. https://doi.org/10.1080/01459740.2010.501313

Wyns, C., De Geyter, C., Calhaz-Jorge, C., Kupka, M. S., Motrenko, T., Smeenk, J., Bergh, C., Tandler-Schneider, A., Rugescu, I. A., & Goossens, V. (2022). ART in Europe, 2018: Results generated from European registries by ESHRE. *Human Reproduction Open, 2022*(3), 1–20. https://doi.org/10.1093/hropen/hoac022

Zafra, I. (2017, February 16). El IVI se fusiona con una empresa de EE UU y crea el mayor grupo de reproducción asistida del mundo. *El País*. https://elpais.com/economia/2017/02/16/actualidad/1487237558_717919.html

Zafra, J. A., Boada Palá, M., Bou Sepúlveda, C., Gómez Cuesta, M. J., Gris Martínez, J. M., & Ramírez López, J. P. (2012). *Recomendaciones para la aplicación del RD 1301/2006*. Sociedad Española de Fertilidad (SEF) and Asociación para el Estudio de la Biología de la Reproducción (ASEBIR).

Zanini, G. (2011). Abandoned by the State, betrayed by the Church: Italian experiences of cross-border reproductive care. *Reproductive Biomedicine Online, 23*(5), 565–572. https://doi.org/10.1016/j.rbmo.2011.08.007

Zanini, G. (2013). *Transnational reproduction: Experiences of Italian reproductive travelers receiving donor gametes and embryos abroad*. PhD Thesis, European University Institute, Florence.

3

Securing a Future

Introduction

Most of my research participants can be described as "clinical laborers" (Cooper & Waldby, 2014) for whom participation in the global reproductive bioeconomy was situated within a framework of instability and precariousness at different levels. They were paradigmatic post-Fordist workers in the sense that their employment options were heavily marked by a context of short-term contracts and self-employment, in which risks are usually outsourced to workers and in which flexibility and the ability to adapt are conditions for employment (Brown, 2015).

Similar to the Indian surrogate mothers in Sharmila Rudrappa's (2015) study, numerous examples from my study show how egg donors drifted between unstable, temporary jobs to egg provision, without a clear perception of the two being any different in terms of conditions. As I show, some even clearly labeled egg donation as a faster, less exhausting, and more meaningful way to make money. Michal Nahman's (2013) examination of Romanian egg donors takes a similar direction when she claims

A. Molas, *Taming Egg Donors*, Health, Technology and Society,
https://doi.org/10.1007/978-981-96-5103-0_3

that the accounts of exploitation and choices "must be taken together with a wider understanding of the kinds of labor (reproductive and otherwise) that are possible and quotidian for people in the context of Romania" (p. 32). Or, as Verena Namberger (2019) puts it when describing the South African egg donation bioeconomy:

> For a lot of donors, it is the everyday reality of a regular 40-hour office job, the boredom of stocking shelves at a supermarket or the hardship of unsatisfying, poorly paid occasional jobs, which shapes their notion of normal labour. In comparison, the duties of being a donor stand out as an exciting and meaningful break from the dull routines of daily life and as a possibility to be "part of something bigger," as many women put it. (p. 99)

What these accounts reveal is the need to understand egg donors' labor lives beyond their passage through the clinic to be able to make sense of their experiences and rationales for undergoing an egg donation cycle.

In this chapter, I address the ways in which participants framed their decision to become egg donors, with a specific focus on its entanglements with the labor context. First, I show how the manufactured dichotomy between the altruistic and the financial interests of egg donors plays out within the speech of the participants. Second, I examine how egg donation and the labor market are intimately intertwined through similar logics, with egg donation serving for some as a way to buy time to look for a job, a job itself, or a way to secure a future and financial autonomy in a context of socioeconomic instability. Third, I explore the rationales of egg donors in deciding whether to complete an additional cycle, and how their circumstances changed their decisions over time. I conclude that undergoing egg donation in Spain is an activity intimately linked to having a job within the post-Fordist context, and sometimes even interchangeable with such a job. As the examples from my research show below, even if egg donation was not considered work by all participants, the close relationship between the labor market and egg donation is present in most of their stories.

"Donated eggs: goods or gifts?" A Critical Approach to Women's Motivations

Although we can discern some patterns, it's difficult to generalize when talking about how donors make sense of their actions and the particular way in which they organized the discourse around presumed altruism and the exchange of money. Nevertheless, most participants were prepared to address the issue of whether their motivations were more "financially motivated" or "altruistic."

In April 2019, when I was 4 months into my fieldwork in Barcelona, I was invited by a Spanish university to take part in a panel on reproductive markets in Madrid. The name of the roundtable discussion in which I participated was called "Gamete donation in Spain: good or gift?" The title, of course, was intended to provoke a lively discussion about gamete configurations in Spain, because it encapsulated the static frame within which the debate has been stuck for years: Are donated eggs goods or gifts?

This is the question with which egg donors are confronted all the time. "So, do they do it for money, or is it really altruistic?," I was asked repeatedly by friends and family during my fieldwork. The answer to this question is presumed to have the capacity to reveal the legitimacy of the whole industry and the true nature of the donor. "Selling yourself for 1000 euros?," someone said to one of my participants, "You don't love yourself." This framework resonates in media approaches to the topic as well as in academia.

In 2014, Guido Pennings et al. (2014) published an article titled "Socio-demographic and fertility-related characteristics and motivations of oocyte donors in eleven European countries." The results indicated that egg donors in Spain were mostly motivated by "a combination of altruistic and financial reasons" (p. 1084), with 56.5% of a sample of 449 donors marking this option (30% indicated "just for altruism" and 19.1% "just for money"). Within a framework that legitimized egg donation to the extent that providers are altruistically motivated, the results of this research were attractive for fertility clinics. As they saw it, most donors actually wanted to help, so we needn't be concerned about a possible exploitation of women or unfair inducement through the economic compensation. In fact, some clinics cited the article to highlight the altruistic motivations of egg donors on their websites targeting recipient couples.

In a critique of this manufactured binary model of thinking, Sara Lafuente-Funes (2017b) argues that focusing on the motivation of women obscures the functioning and profit of the industry in favor of individual moral questions. She notes that looking at egg providers' subjective perceptions in this way places "the motivations of donors at the core of the definition of the process (and not the motivation of the clinics, for instance) and separates the procedure from the idea of the sale of eggs or the exploitation of women"[1] (Lafuente-Funes, 2017a, p. 259). Building on her critique, I contend that asking "Who are the egg donors?" or "What do egg donors think?" from an identity perspective risks exoticizing the women who donate eggs, suggesting that they are somehow different from other young women in Spain.

Similarly, Laura Perler and Carlo Schurr (2020) contend that the financial interest/altruism dichotomy when considering egg donors' motivations is a Eurocentric conceptualization of choice and agency. They argue instead that women's choices must be analyzed "against the backdrop of their [egg donors'] reproductive biographies" (p. 3) to understand the entanglements between the global bioeconomy, state biopolitics, and women's intimate lives. As Andrea Whittaker (2015) puts it, "The critical issue for feminists is not so much the content of women's choices but the conditions under which they are made" (p. 266).

The focus on this dichotomy suppresses important questions about the collective and individual contexts in which these decisions are made. Moving away from static frameworks of motivations, this research aims to uncover the sociotechnical entanglements within which egg donation takes place in Spain. To do this, I focus on egg donors' experiences in the light of discursive, moral, and legislative contexts, in particular, those linked to the Spanish labor market.

[1] Original dissertation in Spanish. Fragment translated into English by the author.

Framing a Wish for Money in Gendered Economies

Among the participants of my research, very few described their donations as *purely altruistic*. Only three out of 25 reported that they had no financial motivation. The main drivers for each of the three were different. Andrea, for example, saw the donation as an opportunity to find out whether she had any fertility problems, as her mother did, while simultaneously helping other women like her mom, who had struggled to have her. Maria, on the other hand, was a midwife who reported that she was "enthusiastic about motherhood-related issues" and said becoming an egg donor had been on her to-do list since she was very young. She donated eggs when she was 32 and chose what for her was the "least commercial place," a private clinic associated with a public hospital in Barcelona. Finally, Stella explained that she donated out of "scientific curiosity" and a desire to help. We met one morning in a cafeteria for the interview, and she did not let me pay for her breakfast, highlighting that she was there to help me with my research, and not to have me "pay for anything." On reflection, I understood that this statement reinforced what she had come to tell me: She did not donate eggs for money, and it was important for her that I understand that. She reported to me that she had initially thought the clinic paid around 500 euros, which she had planned to donate to the non-governmental organization Doctors without Borders. When she found out the payment was actually double that amount, she decided to keep half of it for herself to pay the rent that month.

For the rest of the 22 participants, financial compensation was the driver of their decision to donate. Eleven women told a story where empathy with women was also important, while 10 others only highlighted why they wanted money and how this wish led them to consider egg donation. In the former group, the division between financial motivation and so-called altruistic reasons or the moral part was blurred. In many of the stories, the two fueled each other. As I have also argued elsewhere, for some participants egg donation is seen as a "win-win situation" (Molas & Bestard, 2017), in which their wish for money can be fulfilled in the best possible way, by helping others. Throughout the book,

I examine their perception of both what they gained and the price they paid for it.

Helena, 27 years old, spoke to me for a long time about how her step-grandmother's experience of infertility was an important factor in her decision to donate eggs. However, later in the interview, when asked about how she found out about the possibility of donating, she said:

> So, a friend… I went to see a friend who was ill and so on […] and she told me, "I'm going to do this egg thing since they'll give me cash." And I thought, "What do you mean by 'this egg thing'?" And she told me broadly. Then I… I'm very headstrong. So, I went home, and I started to look it up and look it up and I thought, "Okay, so I'll do it."

Indeed, in numerous stories, participants emphasized the altruism of their donations when I first asked them but went on to reveal strong financial motives as our conversation(s) proceeded. The case of Mica shows the impossibility of grasping the context in which women become egg providers by just asking about their motivation in a single survey question. Had she taken part in Pennings et al.'s (2014) survey, she would have answered that her motivation was both financial and altruistic, and yet that would have told us nothing about Mica or the egg donation arrangements operating in Spain. The story she told me underscores the multiple layers that need to be considered in any study about egg donors' decision-making process:

> I have a friend who has donated several times. Some went well, others worse, and so it scared me a bit. And well, I have relatives, not blood relatives, but who can't… who needed to adopt, who can't have children and all of that. Not here, but… just because of the fact that they couldn't have children… it gave me strength to do it. Because since I can do it, since I can help… And then during the months of August… July, August, September… I had a lot of free time. […] I mean, I'm young. I need 1000 euros. I want to help….
>
> (Mica, 22 years old, two-time egg donor)

Certainly, Mica had free time and could use 1000 euros. Her family migrated from Uruguay to Spain when she was 11 years old and struggled for years to reach financial stability. The family of four lived in a 30-square-meter apartment. Mica's mom worked in a supermarket and her father in a factory, which eventually closed and left him unemployed and depressed. Mica's grades at school were low, and her brother suffered from outbursts of anger. Finally, her father decided to go back to Uruguay, which, as Mica claimed, improved both her grades and her brother's outbursts. However, shortly after that, her mom started a new relationship with a man who was not well received by either of the children, and the conflicts started again. Therefore, when she was 16 years old, Mica decided to return to Uruguay with her father:

> With my father, we were always at the brink of poverty. I mean, it was even hard to buy toilet paper. My dad didn't work so much, but the main thing is that everything there is very expensive. Everything's a sacrifice. It was always rice with mayonnaise... buying a burger was like "Wow!" Everything is uphill. [...] And I would have had to study at the university in Montevideo, which was an hour and a half away, but I couldn't afford the [train] ticket, plus the universities are crowded because they're public, and I said, "Okay, I need to leave."

She came back to Catalonia with her boyfriend from Uruguay, whom she married so he could get the documentation necessary to live and work in Spain. However, her plans to start university in Catalonia proved more difficult to carry out than she had expected. She needed recognition of her studies in Uruguay, which took months to process, and during this time she didn't know whether she would be able to start her studies or miss a year.

Starting in November, she worked full-time as a waitress while she awaited the validation of her Uruguayan secondary-school diploma. Her husband couldn't work because his residency and work papers also took several months to come through. Meanwhile, Mica alone paid rent for both of them to her mother and covered their remaining expenses. The situation worsened when the cafeteria she worked in closed and, by July, she was unemployed, "So the situation was a bit like this. I was going to start studying, Roberto didn't have a job and I didn't work 40 hours anymore, and so the money [for the egg donation] was welcome to pay bills."

Mica's case provides an example of the complex entanglement of experiences, life stories, circumstances, and desires that can lead a woman to consider egg donation. In her case, a migration story, job struggles, and family financial instability all played a role in her decision to become an egg donor. However, Mica didn't revisit the story of her infertile non-blood relatives in any of our remaining conversations, nor in the messages that we exchanged throughout my fieldwork.

Indeed, admitting that they were donating (at least to some extent) for money was not easy for some participants and sometimes caused mixed feelings. This was the case of Júlia, a participant that I met for the first time at the university cafeteria of the School of Psychology. She looked older than the average undergraduate student. She told me that she was 34 years old and that she had recently started university studies for the first time. "I'd always had a very altruistic personality," she said when I asked her about how she decided to donate her eggs. But she also told me that she had her first child when she was 16 and the second when she was 22. She was raising the children on her own, she explained, and she really needed a bit of extra cash.

Júlia already had several strategies to make ends meet, which mostly involved caring for children. She spent an hour a day cleaning her children's school lunchroom, for which she received leftover food in return. As for work, she had several spontaneous temporary jobs that she combined with caring for multiple babies. Laughing, she recalled the time when she was babysitting three babies from different families at the same time while her children were at school. She used to carry the smallest one (4 months old) on her chest, the second one (8 months old) in a baby backpack, and the third one in a stroller. "Tourists took pictures of me," she said, amused. This continued until the night she dreamed she was leaking breastmilk and realized, when she woke up, that she had begun lactating again six years after having stopped breastfeeding her daughter. She then decided to stop babysitting the 4-month-old baby.

Júlia expressed her ambivalence about her presumed altruistic motivation for donating and her need for money in the following way:

> It was a moment that… first, this wish to help that I had, because it's true that I thought about it, right? I had also met people who had difficulties

and who had to turn to this, right? To donor eggs. And well, for me motherhood had always been something that… well, at that time, it was something I was enjoying a lot, right? Like I could empathize, right? […] And there was also a financial need. This was very hard, right? Because during the process, I started to have like a lot of doubts, like "It's not good if I'm doing this for the money," to myself, you know? And "If it's for this financial reason, I don't feel good because I'm not doing this because I want to help, or not entirely," right? And this made me feel very anxious.

An even clearer picture of the uncomfortable feelings surrounding a wish for money emerged from my interview with Sara, who was 25 years old and who had donated her eggs twice. An hour into our interview, she finally told me the reason for her two donations when she was 19. She warned me before starting, "I mean, this is really tough, okay?" and then she told me she used the money from the donations to pay for the orthodontic treatment she needed to stop her debilitating migraines. As she told me, these had started after her family was evicted from their home because her father, who worked in construction, had to shut down his business following the economic crisis in 2008. Again, what started as a story of empathy toward a friend of her aunt who had to undergo IVF to conceive finished with a story of personal hardship involving the loss of a family business, an eviction, and a considerable health care expense.

Mireia, a participant who had donated eggs 5 years before our interview, reflected on her motivation, acknowledging that, when she had claimed to "want to help," what she had really wanted was to help herself:

In a certain way, all the thoughts that I had of "Yes, okay, I want to help someone," it was a bit to comfort myself because I knew that the foundation of my ideas was that I had a financial need. Then yes, all the rest was additional stuff. And it's true that it's something that in that moment made me feel bad, like you're saying "yes," that you think it's really good, that you're doing that for another person, but in reality you're doing this to help yourself, right?

(Mireia, 28 years old, one-time egg donor)

Her experience resonates with Perler and Schurr's (2020) observation that egg donors in Mexico expanded their notion of "solidarity" to include their own personal or family needs, thus allowing egg donation to remain within a logic of care. Similarly, Jociles (2020) suggests that often the motivations of women to donate eggs are related to altruistic actions toward themselves. She argues that the language used by many egg donors to refer to the money received as a form of "help" leaves intact a framework of altruism, although, in this rendering, the "help" is not for egg recipients, but rather for the egg donors themselves.

The case of another participant, Berta, provides a contrasting example. While most donors began with stories in which altruism had a central role, Berta considered herself a victim of the system who had been forced to donate because of her family's financial problems. While Berta's case might seem to contradict the others, its underlying logic is the same in the sense that donors employed epic narratives to explain their choice to donate. In an extract from her diary that she shared with Berta pictured herself as a valiant victim of poverty forced to sell her body to overcome the adversities in an ambivalent narrative in which she loses her principles:

> She goes out on the street early in the morning and thinks about how when she was little, everything was easier, or so she thought, being unaware of all the problems that happened around her and worrying only about asking her mom for 10 euros to go to the movies with her friends.
>
> Now instead she is the one who must bring the money home, and she is totally aware of the things that happen around her. Reality made her grow up all of a sudden with a slap in the face, because when you are lower class, this is what happens. You grow up fast along with the blows life gives you without really being aware of what's happening.
>
> That morning, with that reflection in her mind, she was walking distracted, not paying much attention to the people around her. It wasn't like other mornings. That day she felt totally besieged by her thoughts.
>
> It wasn't the first time she was going to that place [the fertility clinic], but this time her feelings were different. This time they were more visceral and totally opposed to those that accompanied her the first time.
>
> She felt sad, and she had a profound rage inside her. She also felt that her principles got lost on the way to the clinic… but on the other hand, she thought of her family and the immediate need of putting food on the table, which wasn't given by the crumbs of a system and an administration that

oppresses us and that makes us their dependent slaves. That makes us kneel for a meal, a house, or a piece of clothing to keep us from the cold.

She had seen her mom ask for things as basic as assistance to pay for school or food and be denied it. She had seen her mom cry inconsolably so many times out of the rage she felt because of being in that situation, that she had the need and the obligation to do whatever it was to take them out of that position, so that they could live for a while without depending on anyone. And if that meant that she had to sell her body for a few weeks, she would do it, being aware of all the hormonal processes she would have to undertake, the poking, the mood changes, the restrictions... finally, of not being herself during the process and to be the "perfect" woman so everything went well. Not to mention, of course, the emotional process that starts the moment she thinks of taking that path, which does not end in the operating room, but is with her during the following months when she sees a little child, a pregnant woman, or even when she has her period.[2]

(Extract from the diary of Berta, 21 years old, two-time egg donor)

The guilt that both Júlia and Mireia experienced as well as Berta's description of her decision as an act of desperation point to the material consequences of framing egg donation as an act of love and empathy toward others. The oocyte market is "developing in close synergy with pre-existing transnational economies of feminised labour (domestic, sexual and maternal)" (Waldby & Cooper, 2008, p. 64) in which labor is usually difficult to quantify. The idea that reproductive work somehow emanates from a natural predisposition of women toward the care of others has historically made it very difficult to monetize, resulting in different kinds of feminized precarious markets (Lafuente-Funes & Pérez Orozco, 2020)

The fertility clinic as an agent in the making of altruism and, therefore, in the making of gendered expectations of third-party reproductive workers has been described by several researchers (Almeling, 2011; Álvarez Plaza, 2008; Curtis, 2010; Degli Esposti & Pavone, 2019; Hovav, 2019; Jacobson, 2016; Lafuente-Funes, 2017b; Marre et al., 2018; Pande, 2014; Rudrappa & Collins, 2015). However, the examples of Júlia and

[2] Although she is talking about her experience, this fragment is in third person because Berta wrote it this way in the original version in Catalan.

Mireia make clear that the institutional framework of egg donation in Spain does not only constitute a discursive apparatus that assists clinics to create and promote "biodesirable" donors for infertile patients (Gunnarsson Payne, 2015), as well as obedient women (Lafuente-Funes, 2017a; Molas & Perler, 2020; Rivas et al., 2019). These mechanisms also carry embodied consequences for egg donors, which translated into anxiety and guilt when donors didn't comply with the gendered expectation that they should be caring rather than financially motivated.

In the following section, I examine the three basic rationales of the participants for donating eggs: egg donation to buy time to find a job, egg donation as a job, and egg donation to invest in future employment. I argue that, although these rationales are different, they all emanate from the same problem: Spain's precarious and unstable labor market.

Fluidity Between Egg Donation and the Labor Market

Buying Time to Look for a Job

For some participants, becoming an egg donor was not a job itself, but rather a way to obtain extra money while they were looking for one. The decision of Mireia to become an egg donor, for example, was framed by her struggle to find a job to ensure that she could continue her studies. The economic compensation from egg donation substituted the income that formerly came from her job, as she explained in the interview:

> After my first year of university, I worked for a while… I worked here by the coast, I had always worked to pay for my studies, right? So, I finished my first year at university and I had applied for a scholarship,[3] I did the first term of the second year but then… I didn't know yet if… whether I was going to have the scholarship again or not, right? Then I finished the first term in December, and by January, I didn't have money at all to be honest.

---

[3] She is referring to an existing scholarship directed at low-income students to help them pay the fees and that sometimes goes along with a small stipend.

> [...] So, I worked for 6 months in nightclubs and restaurants, making money to go back to university in September and continue my studies. During that summer, one of my coworkers told me that her sister had done an egg donation and that it went very well. She also did it for a matter of money and so on, and it was something that I thought "*hostia*" [damn it]. Also, my sister had my first niece, and it was something that I had never thought about at a personal level, I always thought, "Wow, I don't know if I want to have children" and so on. But with the experience of being an aunt, I had this awakening much more, right? Like, "*Hosti*, I do like children," right? "I do have something there." [...] I had to ask for financial help from my parents, for two months, which I didn't like, and I saw myself quite suffocated and then I... the idea came to my mind again.
> (Mireia, 28 years old, one-time egg donor)

Mireia's goal was to continue her university education, something that, due to her socioeconomic situation, she could only do with external income. The idea of donating eggs attracted her when she was struggling to find a job and needed to ask for financial help from her parents. She also indicated the need to have a moral reason to do it, which in her case was the realization that she actually liked children when she became an aunt.

Another participant, Merche, shows still another way in which egg donation and the precarious labor market are intrinsically linked. As she told me, at 19 years old, after five months of looking for a job "night and day," she decided to schedule a donation cycle "to have a [financial] cushion":

> From June to October, I was looking for a job night and day. I signed up for Infojobs,[4] Adecco, Randstad.[5] I went to the SOC [Public Employment Service] and to a local institution in my town where they help you look for a job and they also do training. [...] I went to business after business with my résumé. I went to Barcelona to get them; I moved heaven and earth. But of course, it reached a point where I said, "Okay, my idea was to start studying, now I can't, I can't save money, I can't do anything. In February,

[4] Employment website.

[5] Temporary work agencies.

> the registration period for the university will start again for next year. I'll have wasted an entire year. I can't find a job," you know? [...] The money [from the donation] wasn't to pay for university, but to have something there just in case, [...] I mean, like it or not, transportation and printing my résumés...
>
> (Merche, 26 years old, one-time egg donor)

For Merche, the donation was related to her work in that it boosted her chances of finding a job and therefore, starting university the next school year. With the money from the donation, she was able to print more résumés, pay for transportation to deliver them, and have a little bit more time to keep looking for employment.

In this sense, egg donation treatments bought time not only for the recipients, who couldn't have children with their own eggs because of their age, but also for egg donors, who attempted to reach their goals with laborious but relatively secure injections of income into their fragile personal economies. Donation provided an amount of money that allowed both Mireia and Merche to be financially independent for some time and aspire to something more than immediate subsistence.

Egg Donation as a *Chosen* Job

A second group of participants conceptualized egg donation and the financial compensation explicitly as a job. Through calculations and comparisons in terms of money and effort, egg provision became preferable to being in the regular labor market. The stories of Martina and Lucía provide a clear picture of the realities of the job opportunities and conditions in Spain and the positioning of young laborers within it.

Martina moved to Barcelona when she was 18 years old. She was starting a degree in Criminology and was leaving her parental home and moving to a big city for the first time in her life. Her parents paid her university fees and her rent in a shared flat:

> My parents are both teachers. I mean, don't think that... [...] it's not so easy for them. I mean, they're making an effort to pay for my education here and if I waste it, I'm very aware that here I'm not on vacation, they'll turn off the tap. I mean, this is very clear.

For the rest of her expenses, she relied on the money she could save up while working as a waitress in Port Aventura, a theme park located in Tarragona (one hour south of Barcelona) in the summer, where she earned 700 euros per month for a full-time contract. However, as she explained, these savings could not cover more than her basic needs in the city:

> Sure, the groceries, the metro for example, the Bicing,[6] I pay all of this myself with the money from the summer. And it covers it, but it doesn't cover, for example, going out every Friday to have a few beers… buying a ticket for a concert, a ticket to the movies twice a month… It's not like I have a very cool social life, don't think that I go out to dinner every night or… partying. It's not my lifestyle, but I mean… with the money from Port Aventura it wasn't enough. […] So… if I hadn't been able to donate eggs, I would have had to accept it. I mean, I would have thought, "Okay, so this is my reality and I have to adapt and accept it as it is," but… there was this option of *doing more things*. I mean… my life didn't depend on going to Bling Bling on a Thursday, to Razzmatazz,[7] to the Love of Lesbian[8] concert… I'm saying things that I've done recently, okay? Things like that. Or to visit my friend who is studying in London. Well, this is already an extra […] All of these are not needs. They were whims or extras that were beyond my summer salary. It didn't cover everything. And since this possibility, [which was] fast and easy, existed, I considered it. And well, it was a bit like that. I mean, damn… I feel like super cold when I say it out loud but… well, I don't think you'll judge me.

Indeed, on one of those early days of her new life in Barcelona, she heard the story of someone who had donated eggs, where she learned for the first time about the money she could get. Since her first donation when she was 20, until her fourth donation, which happened a couple of months before our interview, egg donation had been her primary source of income to cover everything she considered a *capricho* (whim):

[6] Public biking network in Barcelona.
[7] Bling Bling and Razzmatazz are night clubs in Barcelona.
[8] Spanish indie-pop band.

> This has been my salary during these 4 years in Barcelona. [...] I remember that a girl in my class worked at Nostrum[9] and she said, "This is so good, because this way I'll be able to pay for whatever," and I thought "Yes sure, I can also work before Easter for three months... " and then I thought "Oh, no way... I can donate eggs and in one month I have more than two salaries," because adding my salary from Port Aventura of July and August I earn the same as from donating eggs once. So, it's like... I'm sorry to say it like this, but it's an easy way to... [...]. So, if you compare it, I work two months in the summer or during the weekends, or for two weeks I inject these hormones into my body, of course, according to the way I am, my personality, my experience, I *choose* to donate eggs and have this economic compensation.
>
> (Martina, 22 years old, four-time egg donor)

For Martina, egg donation was the job she chose in the light of the possibilities available to her. While she thought that she could work part-time during the week like her friend, she quickly concluded that egg donation gave her a much higher rate of compensation per time spent. I discuss the way she assessed the possible risk of the donation in the following chapter. Her speech also points to the symbolic effect of egg donation as an option available in a young woman's life in Spain. She believed that if it hadn't existed, she would have had to accept her impoverished financial situation. However, egg donation gave her something to aspire to: a more comfortable and fun life. With a notable amount of guilt and shame, she stated that egg donation allowed her to indulge her "whims," which highlights her definition of non-basic items and activities.

In a similar manner, the economic compensation was for Lucía a way to achieve a kind of life that was beyond her reach. When she wanted to move out of her parents' home and move in with her boyfriend, she found in egg donation the job that would allow her to accomplish her goal in the fastest way:

> I was broke, so I started to work as a waitress in a job where they would call me, and I worked by the hour. More like events and so on. So, of course, my pay was very low, and I wanted to have a certain financial indepen-

[9] Take-away food company.

dence, so I didn't have to depend so much on my boyfriend. Because at the end of the day, I was depending on my boyfriend, right? And I didn't... like it very much. So, this way I had a complement of 1000 euros, so I calculated that those 1000 euros would last me more or less for 3 months [...]. Like, I don't know, to be able to go away for the weekend, you know? These things. It was always purely a financial motivation, I tell you. And at that time, it was being able to move out [of my parents' home] [...] I always say that the thing is that it takes me between 10 and 15 days to do all the processes and, in this time, I earn 1000 euros. There is no job where I can get the same. So, basically, for me in these moments, it's like a job. It was like a job.

(Lucía, 22 years old, four-time egg donor)

As Lucía made clear, this group of donors saw egg donation as similar to and even exchangeable with participation in the labor market. This was evident not only in terms of the participants' experiences in both sectors, but also at a material level. As noted earlier, egg donation paid 1000 euros, which is more than what most of the participants could aspire to earn in a month at a full-time job.

Securing Future Employment

Another thing that was often mentioned by women when talking about how they would direct the money earned through egg donation was unpaid internships. Most degrees in Spain require a certain number of practicum hours. Students usually choose among several organizations in their field, some of which pay a small amount of money. However, the most well regarded organizations in a given field are often in high demand among interns and, as a consequence, generally do not offer paid internships. This issue creates inequality between the students who can afford to work for free part-time and attend class the rest of the day (leaving no time for paid employment) and those who cannot.

In order to acquire the more promising internships, some of my research participants decided to complete a donation cycle. That is, paradoxically, they underwent donation to support themselves while performing an unpaid job, in that hope that this would boost their future opportunities. Nancy, for example, had been selected to take part in an exciting internship program in which she would be sent to Portugal for three months. However, there was no stipend:

> I thought that I needed the money, that was it. Because as I was going to start an internship… I'm going for 3 months to do an internship. These are 3 months where I won't earn any money, because this isn't a paid internship. They pay for the travel, the food, the accommodations, but they don't pay for the work itself. I mean, all the rest is paid but the internship is not. So, of course, if I want to spend some money there, I won't have anything. And I don't want to depend on my parents either.
> (Nancy, 22 years old, rejected from the egg donation program)[10]

The situation of Helena was very similar:

> The first time it was more personal [the first cycle], because of what I told you, and the second it was… well, they gave me an economic compensation of 1000 euros and I… I was in the third year of my university degree and my dream was to go to Argentina during the fourth year to do my internship. So it was like, "I'll donate one more time and… and I'll go." […] It was very good, and yes, I was able to leave with a [financial] cushion, that's the truth. Because financially, it's not like I was precisely wealthy, but… if I wanted to achieve this dream, I had to get it from somewhere.
> (Helena, 27 years old, two-time egg donor)

These internships were seen as opening up opportunities in their future. Helena even labeled going to Argentina as her "dream," a dream that required money. Like Merche and Mireia, for whom egg donation was related to their access to university, Nancy and Helena invested in their future careers by funding their period of working for free in exciting environments, experiences that they thought would improve their résumés and therefore their future employability. The language of "dreams" here is interesting, as fertility clinics also employ it to advertise egg donation as a way to make "couples' dreams to make a family come true." In egg donors' narratives we witness a different kind of "dreams" that get animated in fertility clinics, dreams of securing stable futures.

[10] The reasons why she was rejected are detailed in Chap. 5.

As seen in the Introduction and the following two chapters, in the stories of most of my participants, egg donation is tightly intertwined with the labor market. Participants jumped from one to another, sometimes using one to boost the possibilities to get the other or simply interchanging them for a while. The fluidity between the two highlights the precariousness and lack of opportunities in the Spanish labor market, particularly for young women. In a context of embedded and accepted instability, egg donation looked like a good alternative to poorly paid service jobs which, moreover, weren't easy to find.

Lucía, a four-time egg donor, provided a good example of this. While she acknowledged that she would have never done a donation cycle had she had better socioeconomic circumstances, rather than being angry for feeling forced to do it, she was happy that this option existed for people who, like her, needed money:

> Well, I think that I would have liked not to ever have needed the money to the point of having to donate eggs, right? To the point of thinking about the possibility. If I had never needed it, I wouldn't have done it, of course. But I think that given the situation, that you need money, it's a viable option, and that's it… I wouldn't reject it and I don't think it is wrong, neither ethically nor physically. So, I think it's positive that the possibility exists for people like me, who have needed it. However, if I had, I don't know, a basic income, then I wouldn't have done it, right? I mean, if there were other possibilities that would have granted me better financial stability, I wouldn't have done it, of course not.
>
> (Lucía, 22 years old, four-time egg donor)

This example reveals the degree to which participants accepted financial instability; it was embedded in their lives and most didn't question it. Interestingly, we see again here how a participant depicted egg donation not as form of assistance for people who couldn't have children, but rather for women like her, who needed a way to make money in difficult times. In this logic, struggling for money and stability in Spain was seen as normal. As long as their donation cycle wasn't used to cover basic subsistence, participants didn't consider themselves *that poor*.

Money of One's Own

Another desire emanated from the stories of my participants: financial independence. Indeed, egg donors usually expressed their desire to stop depending on other peoples' money. In her diary, Berta explained how fed up she was with depending on what she called "the crumbs of a system and an administration that oppresses us." In a similar manner, both Mireia and Lucía wanted to stop depending on money from their close relationships, the former from her parents and the latter from her boyfriend. In contrast, for Mica and Merche, receiving help wasn't even an option, so they had to make their own "cushions" of money in order to face adversity in uncertain times. It is interesting to observe how intergenerational relationships play out in this choreography. While young women find themselves in this situation in part because of a lack of inherited wealth from previous generations, they ironically become the means of reproduction for older women. The question remains whether they will be able to find some financial stability when they wish to start a family or if it will be too late for them to do so.

The way participants looked to produce some money they could consider "their own" resonates with the problems of women that Viviana Zelizer describes in her book, *The Social Meaning of Money* (1994). As she explains, at the beginning of the twentieth century, American women without access to the labor market used various strategies to gain money of their own. Even the wives of wealthy husbands felt poor in the sense that they not only depended on their husbands financially, but also because, in most cases, they were not even assigned a fixed amount of money from their husbands' salary. Instead, they had to ask for it every time, which meant the continuous monitoring of their expenditures. Zelizer talks about the ways in which women tried to gain some money they could consider their own, such as stealing from their husbands, asking dressmakers, shoemakers, and other sellers to inflate the bill so they could keep the surplus, or even withholding sex until the husband would agree to provide them a regular income. Building on her description, I argue that egg donation in Spain can be understood as a feminine subsistence strategy to meet goals that are mostly directed at achieving a higher degree of personal autonomy.

Going Through Another Cycle?

Nearly all egg donors had to consider whether to undergo another egg donation in the future. This is not because everybody felt eager to donate again, but because if a cycle was successful from the clinic's point of view (that is, if the donor produced between 15 and 25 eggs), the clinic contacted the donor multiple times to inquire whether she would like to donate again. Having to give an answer to the clinic led participants to develop a rationale for why they would or would not donate again. While empathy and solidarity were embedded in most participants' talk when I interviewed them about their first cycle, this theme vanished completely when they considered completing another. On this question, there was a large homogeneity in the donors' responses: They would consider it only if they needed money.

> At the moment, I haven't thought of doing it again. [...] At the end of the day now, I work more, I have some extra income, which isn't 1500 euros, but still.
>
> (Ángela, 22 years old, two-time egg donor)

For Ángela, donating was no longer a necessity now that she was working a bit more and had more income. Similarly, Berta didn't rule out donating again, but would only do so if she had a strong financial need.

> The last time, I also said, "I won't do it again." Right now, I'm at the point of, "I won't do it again," but I don't know how things will go. I don't know. I can't give a yes or a no that is 100% sure, because I don't know. I don't plan to, and maybe at a personal level is not something that I'd say, "Look, I'll go to the clinic and do it," but it also depends on the financial situation I'm in.
>
> (Berta, 21 years old, two-time egg donor)

My experience in the field was that vows not to donate again were frequently broken under further financial pressures. Lucía, for example, said that needing money and the time between donations made her "forget" about the bad experience she had during her second cycle: "It was because of a financial need, and as enough time had passed so as to forget about the pain I had felt the second time… [laughs]. I forgot about the pain and that's it, right? So, I did it."

This "forgetting" was not unique to Lucía. Emily also seemed to experience it when deciding to donate a second time. She was 23 years old when I met her, and she was in her third year of studying anthropology at university. She was doing her undergraduate thesis on mental health and stigma, a topic to which she related personally, as she herself had suffered depression some time ago. She was so passionate about her research that she was considering undertaking a master's degree and maybe even a PhD. We chatted about this possibility, and I told her about my experiences on the journey to a PhD. We met for the first time just one month after she did her first donation cycle, and she was pretty happy about the outcome and keen to do it again:

> So look, I thought it was going to be a much worse experience. It's true that I had some pain and discomfort, but they were pretty… I mean yes, it was hard to do things, but I could do them. So, I don't know, there was a limitation, but well. And then also because… the same as I told you before, I think about the money they will give me, so it's true that sometimes I think that it should be more money, but the money that they give me is already good for me. So, well… since it's not a process that has disrupted my life so much… and the money is good for me, so I focus on that. So, it's very… pragmatic, right?

However, just a couple of weeks after we met, she told me she had not been feeling well. Her level of anxiety was very high, and she had relapsed into depression. She didn't know what was going on but after her counseling session with her psychiatrist, "everything started to make sense." She had had a car accident just before the donation, which she had not had time to process. So, her explanation was that she was suffering from post-traumatic syndrome and that the hormones of the egg donation had further exacerbated her state of mind. She had to stop going to university for several weeks, during which we stayed in touch via text and voice messages. She told me about her slow recovery, and a month later I met her again on campus.

She was looking weak, but she said she was happy to start slowly going back to classes. "I think that the hormones have a lot to do with this," she told me. "I mean, it's a bomb that you put in your body and somehow it has to affect you." To my surprise, however, a few weeks after that second encounter, she texted me to tell me that she had agreed to undergo another cycle at the same clinic. She said she needed the money again, although she didn't tell me exactly what for. She said that this time she knew what the process was, and that therefore she was more prepared for it. I suggested that she might keep a diary to record the different phases of the process. She agreed and later shared it with me. In one of the sections, she wrote this:

> The truth is that the injections are bearable. At the beginning, I don't mind them but as the days pass, my body is heavier and more sensitive. I can't wait for the moment of the extraction. It's hard to walk or to make any physical effort. My belly is bloated, and I almost have to walk with my legs open so that my ovaries don't hurt when I walk. I wonder during the whole process how it's possible that I agreed to do it again. I wonder if sometimes I'm too lax with these things and I start to feel a bit guilty. I think that in part I'm selling myself to the system and the reproduction market, but at the same time, I think what I would do if I wanted to be a mother and I couldn't for whatever reason. I can't avoid thinking that I feel sold, but how important it is for some people to have children. Why is that? Would I be able to continue with my life if they told me that I have problems to conceive or that I won't be able to be a mother? What about the feelings of women who, for whatever reason, socialization, etc., have the profound desire to be mothers? Am I doing the right thing? Am I contributing to a system that exploits reproduction? I have the feeling of taking part in something I don't know a lot of things about, and that provokes a bit of anguish. Not to mention the medication and its effects on the body. Even though I try to deal with it the best I can, it scares me a bit that so many hormones might trigger more anxiety like the last time, even though it's not the same situation.
>
> (Emily, 23 years old, two-time egg donor)

Emily's reflection raises the issues of the labor that donation involves, the lack of information received, her real ability to anticipate risks, donors' understandings of recipients, and the relationship of all of these factors to the legal framework of egg donation in Spain. These topics will be discussed in the next two chapters.

Conclusions

In this chapter, I have discussed the intimate relationship between egg donation and the unstable labor market in Spain. I argue that egg donors in Spain are paradigmatic post-Fordist workers, whose employment options are defined by short-term contracts, self-employment, the assumption of risks, and the requirement of flexibility (Brown, 2015; Cooper & Waldby, 2014).

I contend that the dichotomy "financial/altruistic" does not help us understand the ways egg donors make sense of their motivations. While, for most, the trope of altruism framed their narratives, economic compensation was the foremost driver for them to donate eggs. The dichotomous framework has effects beyond discursive practices and materializes in their stories as feelings of ambivalence, guilt, and shame. It highlights the gendered expectations of women as altruistic rather than financially motivated, which is also promoted by the clinics' discourse.

In examining the stories of my participants with egg donation, I have argued that most of them departed from a common source: the unstable, flexible, and precarious labor market in Spain. In this sense, I have identified three drivers for choosing to donate eggs. The first driver was donating eggs to "buy time" to look for a job. In the two examples selected, participants needed to get a job in order to be able to pay for their studies. Because finding a job was very difficult, after some time they despaired over their lack of money, lack of time, and the barriers these presented to their aspirations of studying at university. In this sense, egg donation helped them get through a stressful time.

Second, some participants highlighted that egg donation was, for them, a job that they preferred over their other possibilities in the labor market. The participants who spoke of this driver highlighted how egg donation provided a relationship between time, effort, and money that no other job could offer them. Moreover, for women in this group, egg donation was a way to upgrade to a slightly more comfortable life in which they could have things other than "basic needs" covered. Egg donation became, in this context, a conscientious "choice" among the range of precarious, lower paid, and more time-consuming labor options.

Third, some participants highlighted that egg donation was a means for them to secure better opportunities for the future. The examples displayed focus on a common phenomenon in Spain: the mandatory internships required for most degrees, which are mostly unpaid. I have argued that young people struggle to juggle them with the studies they need to undertake, because usually this involves not having income at all for several months. However, internships are regarded as important opportunities to enter into the competitive labor market. For this reason, undergoing a donation cycle was a way to support themselves during the time they would be working for free, without any other source of income. It provided a chance to take exciting opportunities directed to securing a better future. Finally, I have pointed out the way in which many of the participants' reasons for donating eggs were related to their aim to become, at least for some time, financially independent from their close relatives or intimate partners. Here I have highlighted the intergenerational choreography in place, in which the lack of wealth transfer from the previous generation ironically led younger women to be the means of reproduction for older women.

In the last section of the chapter, I have examined the rationales of the participants around undergoing or not more donation cycles in the future. Most reported that they would only consider undergoing an additional egg donation cycle if they needed money. Having stable employment and sufficient income to live with some security usually dissipated any interest in egg donation. Yet bad experiences with egg donation were "forgotten" when financial pressure increased, and some donors reversed their decisions not to undergo another donation and ultimately underwent another cycle.

Overall, in this chapter, I have highlighted the entanglements between the precarious labor market in Spain and the participants' decision to donate eggs. I have shown that, even though the gendered framework of altruism permeated the narratives of egg donors, the different strategies through which egg donation was used (buying time, upgrading life possibilities, and securing a future and financial independence) showed the entanglement between economic conditions and aspirations in women's egg donation stories. These results show how egg donation in Spain can be understood as yet another form of self-entrepreneurship with a solid labor structure defined by its fragmentation and insecurity, and in which all the risks are outsourced to the workers. As Charlotte Kroløkke (2015) puts it, donors become "eggpreneurs" who "negotiate and engage neoliberal ideologies, framing egg donation within a moral economy of gifting while also engaging affective economies" (p. 17).

In the following chapter, I discuss in detail the problems related to the availability and quality of information on egg donation and its relationship with temporalities within fertility clinic practice. I explore how egg donors' (in)ability to anticipate risks, their positioning within an anonymous framework, their age, and their precarious contexts orient them toward short-term monetary gains over long-term risk assessment.

References

Almeling, R. (2011). *Sex cells: The medical market for eggs and sperm*. University of California Press.

Álvarez Plaza, C. (2008). *La búsqueda de la eterna fertilidad. Mercantilismo y altruismo en la donación de semen y óvulos*. Alcalá Grupo Editorial y Distribuidor de Libros.

Brown, W. (2015). *Undoing the demos: Neoliberalism's stealth revolution*. Zone Books, Princeton University Press.

Cooper, M., & Waldby, C. (2014). *Clinical labor: Tissue donors and research subjects in the global bioeconomy*. Duke University Press.

Curtis, A. (2010). Giving 'til it hurts: Egg donation and the costs of altruism. *Feminist Formations, 22*(2), 80–100. https://doi.org/10.1353/ff.2010.0009

Degli Esposti, S., & Pavone, V. (2019). Oocyte provision as a (quasi) social market: Insights from Spain. *Social Science & Medicine, 234*, 1–9. https://doi.org/10.1016/j.socscimed.2019.112381

Gunnarsson Payne, J. (2015). Reproduction in transition: Cross-border egg donation, biodesirability and new reproductive subjectivities on the European fertility market. *Gender, Place and Culture, 22*(1), 107–122. https://doi.org/10.1080/0966369X.2013.832656

Hovav, A. (2019). Producing moral palatability in the Mexican surrogacy market. *Gender & Society, 33*(2), 273–295. https://doi.org/10.1177/0891243218823344

Jacobson, H. (2016). *Labor of love: Gestational surrogacy and the work of making babies*. Rutgers University Press.

Jociles Rubio, M. I. (2020). "Supongo que si hubiera tenido pasta, no lo habría hecho": motivaciones para donar óvulos e ideología del altruismo. In A. M. Rivas & C. Álvarez Plaza (Eds.), *Etnografía de los mercados reproductivos: Actores, instituciones y legislaciones* (pp. 35–93) Tirant lo Blanch.

Kroløkke, C. (2015). Have eggs, will travel: The experiences and ethics of global egg donation. *Somatechnics, 5*(1), 12–31. https://doi.org/10.3366/soma.2015.0145

Lafuente-Funes, S. (2017a). *Bioeconomías reproductivas: Los óvulos en la biología pos fecundacion in vitro*. PhD Thesis, Universidad Complutense de Madrid, Madrid.

Lafuente-Funes, S. (2017b). Egg donation in the making: Gender, selection and (in)visibilities in the Spanish bioeconomy of reproduction. In V. Pavone & J. Goven (Eds.), *Bioeconomies* (pp. 253–277). Palgrave Macmillan. https://doi.org/10.1007/978-3-319-55651-2_11

Lafuente-Funes, S., and Pérez Orozco, A. (2020). On (global) care chains in times of crisis: Egg donation and domestic work in Spain. *: Latin American Science, Technology and Society, 3*(1), 354-376. doi:https://doi.org/10.1080/25729861.2020.1796326

Marre, D., San Román, B., & Guerra, D. (2018). On reproductive work in Spain: Transnational adoption, egg donation, surrogacy. *Medical Anthropology, 37*(2), 158–173. https://doi.org/10.1080/01459740.2017.1361947

Molas, A., & Bestard, J. (2017). En Espagne, le don d'ovules entre intérêt, solidarité et précarité. *Ethnologie française, 167*(3), 491–498. https://doi.org/10.3917/ethn.173.0491

Molas, A., and Perler, L. (2020). Selecting women, taming bodies? Body ontologies in egg donation practices in Spain. *: Latin American Science, Technology and Society, 3*(1), 396-414. doi:https://doi.org/10.1080/25729861.2020.1781371

Nahman, M. (2013). *Extractions: An ethnography of reproductive tourism*. Palgrave Macmillan.

Namberger, V. (2019). *The reproductive body at work: The South African bioeconomy of egg donation*. Routledge.

Pande, A. (2014). *Wombs in labor: Transnational commercial surrogacy in India*. Columbia University Press.
Pennings, G., de Mouzon, J., Shenfield, F., Ferraretti, A. P., Mardesic, T., Ruiz, A., & Goossens, V. (2014). Socio-demographic and fertility-related characteristics and motivations of oocyte donors in eleven European countries. *Human Reproduction, 29*(5), 1076–1089. https://doi.org/10.1093/humrep/deu048
Perler, L., & Schurr, C. (2020). Intimate lives in the global bioeconomy: Reproductive biographies of Mexican egg donors. *Body & Society, 27*, 3–27. https://doi.org/10.1177/1357034X2093632
Rivas, A. M., Lores, F., & Jociles, M. (2019). El anonimato y el altruismo en la donación de gametos: la producción de biocapital en la industria reproductiva. *Política y sociedad, 56*(3), 623–644. https://doi.org/10.5209/poso.60564
Rudrappa, S. (2015). *Discounted life: The price of global surrogacy*. New York University Press.
Rudrappa, S., & Collins, C. (2015). Altruistic agencies and compassionate consumers: Moral framing of transnational surrogacy. *Gender & Society, 29*(6), 937–959. https://doi.org/10.1177/0891243215602922
Waldby, C., & Cooper, M. (2008). The biopolitics of reproduction: Post-Fordist biotechnology and women's clinical labour. *Australian Feminist Studies, 23*(55), 57–73. https://doi.org/10.1080/08164640701816223
Whittaker, A. (2015). Technology, biopolitics, rationalities and choices: Recent studies of reproduction. *Medical Anthropology, 34*(3), 259–273. https://doi.org/10.1080/01459740.2015.1019066
Zelizer, V. (1994). *The social meaning of money*. Princeton University Press.

4

Tempting Luck

Introduction

Well, we asked our friend what we had to do, what steps we had to follow and so on. […] And well, that's it, we started to talk. I said "Well, so let's do it." I have nothing to lose, I guess, one more experience. […] Plus, it was because I wanted to continue my studies, so I said, "Let's go, why not?"
(Laia, 26 years old, two-time egg donor)

At the end of the day, I don't know, it's for science, or to help people or whatever. So, if it doesn't involve any risks for you, why not?
(Stella, 20 years old, one-time egg donor)

One of the most intriguing aspects of the fieldwork for this research was the range of different perceptions about risks across participants. While some claimed to fear the risks of egg donation, very few were able to name any specific ones. Others, like Laia and Stella couldn't even see a reason for thinking twice about undergoing an egg donation cycle. "Why not?" they rhetorically asked. Why not?

Contemporary debates on egg donation arrangements around the world increasingly focus on questions of information delivery, informed consent, and comprehension. These discussions point out the existing

A. Molas, *Taming Egg Donors*, Health, Technology and Society,
https://doi.org/10.1007/978-981-96-5103-0_4

problems in providing informed consent for egg donation when the long-term risks remain unknown (Beeson & Lippman, 2006; Brody, 2017; Pearson, 2006; Schneider, 2008; Schneider et al., 2017; Tober et al., 2021; Woodriff et al., 2014). In this chapter, I add to these debates by shedding light on the information provided by Spanish fertility clinics to egg donors, the reception of this information by the participants, and the resulting understandings it produced. I argue that the production of these understandings relates to the "temporal choreography" in fertility clinics, which is part of the taming process of egg donors.

Charis Thompson notes the intricate coordination involved in assisted reproduction, which she refers to as an "ontological choreography." In her definition, this choreography encompasses the dynamic synchronization of various elements within ART clinics, including technical, scientific, kinship, gender, emotional, legal, political, and financial aspects (2005, p. 8). Thompson emphasizes clinics' role in merging entities that are traditionally viewed as belonging to separate ontological realms, such as nature, the self, and society (2005, p. 8). Expanding upon Thompson's influential observation, in this chapter I explore how different temporalities are coordinated within the practice of ART clinics. I propose the term "temporal choreography" to describe this coordination of social, biographical, and biological relationships with time in the context of assisted reproduction (Molas & Whittaker, 2023).

I focus on the role of temporalities in shaping the provision, reception, and perception of information, as well as their impact on donors' ability to anticipate and manage risks. I argue that within the clinics, practices of delivering information for informed consent framed risks within the present, highlighting the immediate consequences of donation, while not mentioning the lack of long-term studies on physical and psychological risks. The legal framework in Spain, in which the anonymity of gamete donation is mandatory, further contributes to donors being inadequately informed about the potential future outcomes of their cycles. Finally, I argue that the biographical temporality of donors, characterized by their young age and often associated with self-reported "emotional immaturity" and financial insecurity in their current lives, shaped their level of engagement and anticipation regarding risks. All these factors collectively placed egg donors in a state of "temporal stasis" that focused on the immediate present or very near future.

By concentrating on the interplay between the temporalities of egg donors and the needs of the Spanish egg donation industry, I argue that these arrangements minimized donors' perception of risks, delimiting their participation and knowledge, dissociating them from the value of their own contribution, and using their eggs as bio-objects owned by the clinic, whose future would depend on market dynamics. These results are particularly significant when we consider that the fertility industry is increasingly being studied within the context of an ethos of anticipation as a form of self-care (Adams et al., 2009). This raises important questions regarding which bodies are encouraged to anticipate future risks such as infertility, and which bodies are compelled to restrict and circumscribe their temporality to serve the demands of the fertility industry.

Clinical Practices: Limiting Risks in the Present

The first way egg donors were circumscribed in time is in relation to the information they received about egg donation. Fertility clinics invest significant effort in building their corporate identity on the websites directed to egg donors (Molas & Whittaker, 2021). While many of them offer multiple tabs, big colorful pictures, and messages highlighting the presumed progressive spirit and empowerment of women who decide to donate eggs, detailed information about the medical process, the risks, and the side effects it implies is harder to find. Instead, clinics regularly update their blogs and social media profiles directed to egg donors, through which "websites create a complete sociotechnical imaginary in which to donate eggs would not just be an act of solidarity but would also imply belonging to a community of progressive empowered women, in control over their bodies, with modern consumption capacity" (Molas & Whittaker, 2021, p. 342).

Once women approached a clinic, they received further information so that they could be asked to give their informed consent. According to my field observations and interviews with both egg donors and clinicians, while the way women were informed about egg donation varied slightly from clinic to clinic, the overall logics shared many aspects with those found online, which not only downplayed risks, but also circumscribed them into the present.

One day, during my observations at the reception desk of the clinic, it was around 5:30 pm when a young woman arrived for her first appointment. She was a friend of a "very good donor," Núria (nurse) had told me before. The woman, Catalina, was from Argentina and had arrived in Barcelona only two months earlier. After a brief oral questionnaire about her phenotypical features and those of her family, we soon moved from the reception desk to a consultation room for the first informational interview.

I sat on one side of the table on a stool next to Núria. Catalina was in front of us. With a PowerPoint that contained only images, Núria started explaining that egg donation was needed by "lots of different kinds of women." Representing the donors was a group of women of different phenotypes who had their thumbs up and were smiling. Next to them were images of other women: some with a man, representing a male partner and some with another woman, representing a female partner. Catalina rested her elbow on the table, propping her head up with her hand, and nodded in silence. Núria said that egg donation was a "voluntary act," which meant that "you don't come here because someone is making you do it; this is altruistic." But since it was a process that "had costs," Núria continued, such as attending appointments, injecting the medication, and undergoing surgery, there was an economic compensation that at the clinic was 1100 euros. Núria highlighted that this was given if "everything goes alright," and then she specified, "We understand 'everything goes right' as: coming to the appointments, taking the medications, etc."

She also told Catalina that the donation would not have consequences for her future fertility, because "we are going to take what would be disposed of anyway." As for medical tests, she said, Catalina would have to undergo a blood test and a genetic test to find out if she was carrying genetic diseases. "This is not a test that people would usually do," said Núria almost proudly. Núria also mentioned the egg extraction, pointing out that it was "simple and short."

About the risks and side effects, Núria said that they were not extraordinary: "bloating like in a period, but stronger." She rapidly mentioned the most common complications, specifying ovarian hyperstimulation (and noted that this had never happened at the clinic) and ovarian torsion. With a reassuring look, she told Catalina that if she followed the instructions from the beginning, there shouldn't be any problem. The last slide of the presentation was titled, "FAQs for interested women." The questions were about the supposed link between egg donation and fertility problems, weight gain, and acne. All of them, she said, are myths. Moreover, she assured Catalina, "This does not affect you at all when you want to be a mother." "At all?" asked Catalina. "No," said Núria.

I had long wanted to be present at an informational appointment for intending egg donors, and the meeting left me shocked by the lack of details in the information provided. This was the only specific appointment intending egg donors had to be informed about the process, its implications, and its purpose. Yet it had only briefly focused on the steps of the treatment and the operation, the conditions in which Catalina would receive the compensation, and reassurances about the unlikelihood of anything going wrong.

In the end, Catalina was not accepted into the program because she had been in Latin America during the last 3 months, meaning that she was at risk for having contracted Hepatitis B. If she were infected, the disease could be transmitted through her eggs. Had Catalina passed her screening, however, the next step would have been for her to read and complete several forms, including the informed consent.

In the context of the proliferation and growth of bioeconomies, Nikolas Rose (2007) asserts that "as biotech companies seek to commodify products—DNA sequences, tissues, stem cells, organs—it is clear that ethics has a crucial function in market creation. Products that do not come with appropriate ethical guarantees, notably assurances as to the 'informed consent' of donors, will not find it easy to travel around the circuits of biocapital. It is also clear that the routinization of ethical concerns in the bureaucratic procedures of research governance can serve to insulate researchers rather than to constrain them" (p. 30). In the context of egg donation in Spain, too, there is an instrumentalization of informed consent forms as bureaucratic tools to assist the commercialization of bio-products rather than an asset to protect rights and address vulnerabilities. As Waldby and Mitchell (2006) claim, informed consent "is the mechanism that transforms a gift into property" and "serves to regulate and formalize the transfer of possession from donor to recipient" (p. 71). This is a process that Verena Namberger (2019) labels a "contemporary form of dispossession" (p. 110) as it alienates the provider from her genetic/biomaterial. All these tensions show "the double face of bureaucracy, as a form of domination and oppression as well as of protection and liberation, and all the ambivalences this dialectic entails" (Bierschenk, 2019, p. 117).

The informed consent forms for egg donation I had access to during my fieldwork followed the exact or very similar format of the recommended template proposed by the Spanish Fertility Association. The risks stated in the document included ovarian hyperstimulation syndrome, intolerance or side effects due to the medication, abdominoperineal infection, hemorrhage due to an accidental puncture of the blood vessels; abdominal pain, accidental puncture in another part of the body, and ovarian torsion. Like what I observed in the description of the risks in the online information about egg donation, the informed consent forms emphasized the unlikelihood of these problems occurring. Clinics did not mention the lack of studies about long-term risks, nor the potential risks highlighted in case study literature. The risks were therefore presented as delimited in the present.

"She doesn't have any clout": Alienating Donors from Commercial Futures

A further practice that delimited the temporality of the act of donation was that of restricting information. So far, I have discussed the ethics of information delivery in egg donation considering only the information about risks and side effects. However, given the transnational dimension of the lucrative Spanish fertility industry, the question remains: What should be considered "relevant information" for intending egg donors? Given the commercial arrangements in which egg donation happens, I contend that not only the risk and side effects, but also the possible futures of their eggs and the exchange value they will acquire once they are outside their bodies are crucial questions to disclose if we aim for "informed decisions" (see Cattapan, 2016; Niemiec & Howard, 2020; Rivas et al., 2019; Tober et al., 2021).

This omission of information is propelled by both the altruistic and the anonymous framework of egg donation in Spain, which results in the alienation of egg donors from the value of their product and helps maintain low production prices. In other words, the imaginaries of egg donation, managed and promoted by the clinic, play a role in the functioning of the industry. As Rivas et al. (2019) argue:

> Anonymity as information control has been extended not only to the non-identifying data of the donor, but also to other data about which the law is silent or ambiguous, and that information is handled according to the clinics' decisions. For example, the information about the number of eggs extracted, their use for one or several recipients, whether they are used fresh or in cryopreservation, the number of pre-embryos obtained after the insemination of the eggs, the success or failure of the transfer, the number of births nationally or abroad, [or] the destiny of the unassigned eggs/embryos.[1] (p. 634)

Confirming the concerns this quotation raises, the participants of this research had little information about the outcomes of their cycles and the market dynamics of fertility clinics. This sometimes made interviews delicate conversations in which I, as a researcher with privileged information, had to decide whether to intervene when women expressed incorrect assertions about their donation cycle. This occurred when I asked participants: How do you imagine the recipient(s) of your eggs? Have you ever thought of the possibility of there being more than one recipient of your donation? Have you ever considered the possibility of the recipients being from other countries? The responses to these questions provided very different answers, which highlighted the fragmentation of knowledge among women donating eggs:

[1] Translated from the original Spanish by the author.

They told me that more or less they would take between three and four eggs. Then, the rest were left there. And then, in the lab, they would check which one was the best to make a… Before that I had seen a documentary where the woman would get the most ideal egg and the rest, in some places, they would freeze them, and in others, they would get rid of them. So, of course, I don't know exactly what they did with the rest, I don't even know how many they took. I know that it was between three and four… and that they chose the most ideal and so on. And I suppose that the rest have been frozen or thrown away. I don't exactly know what happened with the rest of the eggs.

(Stella, 20 years old, one-time egg donor)

Stella, who like numerous other participants reported not having been given information about the outcomes of the cycle, did not understand the number of eggs that are usually extracted in a cycle in Spain (around 20, see Chap. 2) or the possible future uses of the eggs, which, if not used fresh after the extraction, would be frozen for future use usually in different recipients. We see similar information gaps in other participants:

Anna: Did they tell you if your eggs were going to be for one or more recipients?

Emily: No. But I don't know if more women can use them. I think there is a law that says that you can't… I mean… or can you? […] In theory… well they didn't say anything in relation to this. And in the documents that I read it didn't say it either. They talked about *the* recipient; I know that.

(Emily, 23 years old, two-time egg donor)

Anna: What do you think has happened to your eggs? Did you ever consider that they may have been assigned to a foreign recipient?

Lídia: […] I had never imagined, never ever, that he [the baby] could be in another country. Not this.

(Lídia, 27 years old, one-time egg donor)

Anna: Have you always thought your eggs were in Spain?

Lucía: No. No, because in the clinic where I went, almost all the people who were there, all the couples who came to receive [eggs], they were almost all tourists. I think it was people who came specifically to do it, the *in vitro* fertilization. In fact, the girls who were at the reception desk, which were four, they spoke like six or seven languages each. Because I saw that they were talking to a lot of people. I mean, that's super international, so I imagine him [the baby] from any place in the world. So, the probabilities that I see one of these babies of mine… [laughs].

(Lucía, 22 years old, four-time egg donor)

Anna: Have you ever thought of the possibility that the couple could be from another country?

Sonia: No, no! [laughs]. Now that you say it, yeah, yeah, it could be, of course, but no, no. I mean, it was not the image that came to mind. I had the image of someone from here, from Barcelona.

(Sonia, 31 years old, one-time egg donor)

As the examples demonstrate, the imaginaries in relation to the future of their eggs varied a lot from donor to donor. However, these different imaginaries usually were not related to the information they were given at the clinic, but rather to the information they had gathered through observation during their appointments or, on some occasions, the information they found in the media.

Most of the participants were not told the number of eggs that had been extracted, the number of couples that might receive them, their recipients' country of origin, the reasons for the recipients to seek eggs, or whether their eggs would be frozen to be shipped abroad—options that most donors did not even know existed. The most common imaginary was a single baby born to a couple that lived nearby and for whom, for medical reasons, using donated eggs was their only chance to become pregnant. In the following section, I shed light on the perspectives and rationalities of clinicians when providing information about egg donation.

The interviews with clinicians in relation to informational practices aligned very well with what I found by talking to egg donors. They indeed highlighted that detailed information of the outcomes of the donation was not communicated. However, they sometimes let donors know if the donation had *led to pregnancy*. As Aina and Neus, gynecologists, explained, while the number of eggs the clinic finally extracted was not shared, donors were instead encouraged to ask *if they helped*. This practice reinforced a narrative centered not only on altruism, but also on the imaginary of the single family they were assisting.

Anna: What information can a donor have? I mean, can she know if the treatment worked? How many eggs were extracted?

Neus: If they ask for it, "Did I help?" we tell them if they led to a pregnancy or not, because this gives them information about their own fertility. [...] The number of eggs by default is not disclosed, but some of them ask "And how many follicles... how many little circles do I have?" and sometimes you tell them, "Look, you have very healthy ovaries" or "We're a bit at the limit, but we'll see how it goes...." [meaning that the patient's ovaries had produced fewer than the expected number of ova]. Usually, they aren't given exact data, but they're told, "Look, your ovaries have a big reserve," or, "Not so much." Obviously if they have a low reserve, we tell them.

(Neus, gynecologist and clinic manager)

Anna: When you're doing check-ups with the donors and so on, what can the donor know about the process she's going through?

Aina: If her donation has led to gestations. If the donor asks, we can tell her. It's like... especially the ones who do it... in an altruistic way, they are interested, and they ask you, "Do you know if the recipient was able to get pregnant?" And you can tell them this.

Anna: And what about more technical things such as how many follicles she is producing and so on?

Aina: How many eggs, right?

Anna: Yes. Would you say it in an ultrasound check-up for example?

Aina: Oh, in a check-up they see it, because they can see it on the screen and they say "Oh, so many!" and so on, and a lot ask about it and you go explaining how the stimulation is going, poor things. There are some who are educated, in the sense that they have... that they are studying, we even have medical students, and you explain it. And they love it. What is not really told is how many eggs were extracted because... At least here they [the management] don't like it much for us to say it. I don't know why. I guess not to give so much detailed information... At the end of the day, she *ni pincha ni corta* [doesn't have any clout].

(Aina, gynecologist)

By using the Spanish expression "*ni pincha ni corta*" ("[she] doesn't have any clout"), Aina expressed that egg donors' participation and relation to her eggs is finished right after the extraction. Literally translated as "[she] neither pricks nor cuts," the idiom is used to highlight that something or someone does not have any power or influence regarding an issue. In this context, Aina stressed that donors do not have sociotechnical power or influence in the decision-making surrounding the destiny of their eggs after the extraction, for which giving them information appears pointless. The number of eggs extracted was considered information that was no longer her business and was now a commercial secret. However, whether someone got pregnant using her eggs was information that she was encouraged to seek. I argue that the obfuscation prevents debates on redistributive justice and maintains the roles of the participants. Letting her know if a pregnancy took place—but not the number of eggs extracted—reaffirmed the gendered logic of altruism and traced the limits of her appropriate interest in her donation, a crucial step in the taming process of egg donors.

Often, the logics of clinicians not to provide information about the outcome of the donation were related to the legal framework of anonymity. Indeed, the fact that donors could not have information about the recipient couples or the resulting children of their donations served as a pretext to obscure all information related to the outcome of the cycle. Nevertheless, other rationalities emerged during the fieldwork. Virginia, egg donation program coordinator, reflected on the vulnerabilities she perceived could emerge by giving a donor information on the results of her cycle. According to her, the information could negatively affect her future reproductive life.

Virginia: For example, if you reject her, you have to tell her the reason, okay? However, are you obligated to tell her how many eggs you have obtained or if the patient is pregnant or not? No, this is not written anywhere. It seems a bit counter-productive to give this information. I think it's information that... it's not because I want to oppose transparency, since I'm the kind [of person] that thinks that everything should be transparent. Everything, okay? I could even get behind non-anonymity. Although I think this is a society that isn't ready for non-anonymity and moreover... from the business point of view everything would sink, okay? But I'm not speaking as a business; I'm talking about my convictions which do not necessarily fit with... [laughs]

Anna: With the business goals?

> **Virginia:** Exactly. These are two different things, and I work, and I earn a salary… like everyone else, right? But my convictions can go in another direction. Okay, so getting back to the point, although I stand for transparency, there is a point that I don't see quite clearly about a donor knowing everything that's happening. I'll tell you why, because if you're a young girl, and you've donated, and I tell you that we didn't get a pregnancy, maybe this will block you in terms of your future motherhood plans. You can think that it won't work, that your eggs don't work… okay? Psychologically this can affect you. And if I tell you that it went alright, it can also affect you psychologically. It can affect you because when you have children you can get obsessed thinking that you have stepsiblings, that your children have stepsiblings around the world, okay? Around the whole world moreover [laughs]. Or it can affect you if you can't have children and you know that there are children who are biologically yours. Okay?
>
> (Virginia, biologist and egg donation program coordinator)

Virginia's reflection brings to the table a topic that needs further reflection: how information about an egg donor's cycle can affect her future. As I show throughout this book, this debate is not exclusive to the outcomes of the cycle itself (i.e., how many eggs extracted or how many children born), but also pertains to the outcomes of the different tests she will undergo as part of the screening process, notably the anti-Mullerian hormone test[2] and the genetic carrier test (explored in Chap. 7). I argue, however, that information about the functioning of the egg donation industry should not be considered sensitive for egg donors. In a rapidly shifting context, it is crucial that egg donors have information on value exchange and the possible futures of donated eggs. Providing this information would help deindividualize the perspective of egg donors and allow informed reflections at a collective level.

I contend, therefore, that there are at least four debates intertwined in this discussion: The first has to do with the quality of information delivery practices. The second has to do with the ethics of disclosing personal medical information to egg donors. The third is related to the debate

[2] Test currently performed on all intending donors as part of the screening process. Anti-Mullerian Hormone (AMH) is a hormone secreted by cells in developing follicles. The level of AMH in a woman's blood is generally a good indicator of her ovarian reserve.

around the anonymity of egg donors. And the fourth involves the ethics of disclosing information about the egg donation bioeconomy, especially when in Spain donation is framed as an altruistic act.

While the debate concerning the anonymity of donation in Spain has been on the table for years (see Chap. 2), the debates on the quality of the information provided, the consequences for the donor due to information being disclosed or withheld, and whether information regarding the business model of donation is required as part of informed choice need to have their own space. At present, secrecy and avoidance stand in the place of the four debates, ultimately benefiting clinics' business goals.

Circumscribing Consent in Time and Place

Temporal circumscription was also present in the way information was given (usually very fast) and the way the different documents were signed (usually inside the clinic during the first appointment, without the option of taking them home before signing). For those who passed the initial screening, the next stage involved an appointment at which the donor filled in and signed multiple documents including a data protection informed consent form (1 page), the egg donation informed consent form (8 pages), a questionnaire about the woman's demographic and medical history (4 pages), the genetic testing informed consent form (7 pages), and the psychological screening test (a multiple-choice questionnaire of 144 questions).[3] The completion and signing of these documents was usually followed by a first blood test or ultrasound scan, all in the same day.

According to my observations at the clinic and to the experiences of most of my participants, egg donors were usually not allowed to take the numerous consent forms and questionnaires home for reading, considering, and signing. Rather the clinic compressed the processes of deliberation and consent into a specific, narrow time-space, usually the first appointment.

[3] Data from one fertility clinic in Barcelona; there might be slight variations across clinics.

> The first appointment with the doctor, I remember that he was a man… not really friendly and it was… it made me doubt [going forward] a lot. I don't know, he was explaining it as if everything was very beautiful, and you always need to watch out, right? […] He was a very commercial man, and everything was perfect, everything was beautiful, and everything you were going to do there was great. So, when I started to ask questions, the details, right? Like "Yes, but this has risks," he was like, "Yes, but don't worry, whatever," and he made me… I mean, he made you sign the documents in the same consultation room. I told him no, that I wanted to take them home to sign them there. And he said no that I had to sign them there because they were the documents for the procedure […]. And I told him, "Okay, so no problem, I'll stay here" [to read them] and I stayed like two entire hours.
> (Andrea, 29 years old, two-time egg donor)

Like Andrea, Nancy also highlighted the number of documents she was required to read and sign in the waiting room the first day she stepped into the clinic. In her case, she didn't even have a "first interview" before reading and signing the informed consent forms:

> I mean, they came to me and gave me like a folder with a lot of documents and told me about it broadly… Well, she told me that I had to read all of that and that she had marked the boxes where I had to sign. But of course, I had to read all that information on my own. I was there a long time reading it all and so on.
> (Nancy, 22 years old, rejected from the donation program)

The multiple documents intending egg donors received at their first appointment need to be analyzed as objects that carry symbolic value by materializing the relationship between the donor and the clinic. While, on the one hand, they served to foster transparency, on the other, they were also material manifestations of biomedical power and authority which in turn became a "reference for truth or authenticity" (Borrelli & Lindberg, 2019, p. 52). These documents did not "reflect" a reality, but rather they produced an official one that looked objective and devoid of human participation (Bierschenk, 2019). By condensing the time required for reading and completing these multiple screening and legal documents into a single appointment, donors had their decision-making process and opportunities for reflection and questions compressed and time limited. Receiving an ultrasound and screening blood tests immediately after signing the papers overshadowed any potential doubts and created a further commitment, as women were scheduled for an appointment to receive the results (see Chap. 6).

Biographical Temporality: Age, Cultural Capital, and Risk Perception

While Andrea and Nancy assured me that they had insisted on reading and understanding all the documents before signing, this was not the case for everybody. Several other examples highlight how the structural and specific temporalities of egg donors influenced their reception of information and reflect the need to understand women's decision-making within a broader biographical frame.

> They told me the pros and cons and so on, but of course, I was 19 years old. As long as they didn't say that I was going to die… you know?
>
> (Alba, 25 years old, one-time egg donor)
>
> In that moment, I experienced it in a… [she exhales] in a very dispassionate way to be honest. I mean, I think they gave me a pen drive with a PowerPoint, but I remember that I didn't pay attention to it at all, I mean… I was in a situation when I was with a partner that abused me psychologically. So it was, "I'm doing this and that's it." And I don't know if they called me and I said, "Yes, yes, everything is fine," like very impersonal from my part, you know? Then I experienced it like, "Let's do it, let's do it," and I was totally unaware. And even the day they did the procedure, I put that injection in the wrong way, and it was a big drama. Everything was full of blood, and so on but I mean, I was in that mood, you know?
>
> (Lídia, 27 years old, one-time egg donor)
>
> But at the end of the day, you have some money on the table, you have a contract that says that yes, you might have a hemorrhage and so on, but then the doctor is telling you that this happens once in so many cases. Of course, you see that the risk is very small, that you have money and it's like, "I don't care what you tell me, I'll sign it and I'll do it." Then here I think

> that there's a part... the ethical part, right? Because when you show money to someone that needs it... how many people sell their body out of need and not for pleasure?
> (Andrea, 29 years old, two-time egg donor)

The experiences and reflections of my participants highlight problems of consent in egg donation that go beyond misinformation. Even receiving clear and complete information about the process and risks, both known and unknown, didn't always change participants' perception of risk. This was the case of Mica. In the two interviews we had, she stated several times that she would not donate again. The problems she had injecting the medication in a train station at night (described in Chap. 5) as well as the physical pain she felt in the days after the donation discouraged her from ever donating again. However, four months after our second interview, she sent me a WhatsApp message. In it, she said that she was thinking of undergoing another cycle and asked me if I could provide her with a list of the "side effects, consequences and so on." She said that she had a new boyfriend and that he wanted to know what the risks of egg donation were. I drafted an email as carefully and clearly as I could. I wrote that the most important thing to understand was that there was very limited data about the long-term risks of egg donation. I recommended to her a *New York Times* article by Jane E. Brody (2017) that had been translated to Spanish, titled "Do Egg Donors Face Long-Term Risks?" which precisely tackled the issue of the lack of studies. Then, I listed the risks stated by the Spanish Fertility Association. I ended the message asking her if she could tell me her final decision and why she was thinking of going through another cycle.

She replied shortly after with a voice message:

> Hey, darling. Well, so now I have a part-time job and I'm starting classes again and so I'm looking for... Well now I'll be hired for a second job so I might not do the egg thing, but I wanted this money that... Well, we know how it is, it's pretty fast, and then it would allow me to have a [financial] cushion for school fees, books, the T-Jove [public transportation ticket], transportation costs... Everything that starting the school year in September involves. So... I thought about doing it and I was telling my new boyfriend

> about it, and well, as usually happens, he didn't really like it. But, he doesn't know about the process in depth, so I told him that I had talked to you and so on, to see if you could send me this list, which has been very useful to know about the side effects and about what could happen or go wrong. And well, he still doesn't agree with me doing it because of the issue of... since they don't quite know in the long-term how it affects [you] so... but I still haven't resolved it. Next week, I'll start at this second job, so I would do it [the donation cycle] in September, October... but well, if I do it, I'll let you know!
> (Mica, 22 years old, two-time egg donor)

At the beginning of October, she wrote to me again to tell me that she had just started a new cycle. Indeed, although having provided her with important information that most donors did not have and having had a tough experience in her first cycle, she still chose to do it one more time. As her voice message implied, the purpose of the list of risks was not to decide for herself, but to convince her partner that he should not worry. Even if my reply probably did not help her in this endeavor, it did not seem to change her will to go on with the process one more time.

While numerous examples highlight how the means and the context in which information about egg donation is given in the Spanish context was problematic, in egg donors' accounts, information was not the only decisive factor in choosing whether to donate. I argue that the debate needs to shift to include the entanglements that co-produce risk perception.

Bioethical literature has focused on the debates around the limited ability of participants to fully understand informed consent forms for biomedical research (Beskow & Weinfurt, 2019), and several authors have attempted to provide guidelines and strategies for the writing and delivery of informed consent information for donors (Beskow et al., 2020; Cattapan, 2016). The largest study to focus on egg donors' perceptions of informed consent in the US concludes that, while most participants felt informed about the short-term risks, more than half of them did not feel well informed about potential long-term risks (Tober et al., 2021). This conclusion aligns with the current debates about the lack of longitudinal studies on the risks of egg donation (Brody, 2017; Coeytaux & Tober, 2013; Woodruff, 2017).

Indeed, individuals rarely have the ability to challenge official discourses promoted by biomedical institutions and supported by the state. As is already implicit in the influential work by Carlos Novas (2006) on patient activism and the political economy of hope, cultural capital is needed to articulate, organize, and coordinate communities to contest biomedical discourses and practices. This was also reflected in my fieldwork, in which participants who were older and had more education were usually the ones who questioned clinics' narratives and had qualms about information arrangements. For example, Andrea put it clearly during her interview. Her account was powerful because she described herself as someone who had socially and culturally "climbed" in recent years. Having quit her studies to work in a call center where she earned 600 euros/month in a full-time contract, she finally decided to start university when she was 26 years old, which she described as a life-changing experience. She highlighted the dangers of deciding to donate when reliable information sources were hard to discern:

> Of course, because according to the information they give you there, everything is very beautiful, like, "No, it's alright, a lot of girls come here, we've never had a case of…" and surely they have had cases of hemorrhage and problems, but they're like, "We've never had a case, whatever…." They downplay the risk really, right? […] I mean, me for example, if I had done the donation when I was 18, I didn't even know that there was such a thing as "reliable sources" […] So, it's this lack of information that I think is really important, because then there are 18-year-old girls whose parents don't give [them] any money, and they go there and maybe it says [in the informed consent form] that you can have an internal hemorrhage during the procedure, but if the doctor, *a doctor*, is telling you, "Don't worry, stay calm, this has never happened here." Maybe they've had 10 cases or more, but what is he going to say?
>
> (Andrea, 29 years old, two-time egg donor)

Andrea highlighted the difficulties of challenging the explanations of doctors and clinic staff as authoritative expert figures and how this made young women who lacked cultural capital particularly vulnerable. Her words resonate with those of Åsa Boholm (2015) when she argues that the definitions, classifications, and management strategies of risks, however, are to a great extent in the hands of "experts." As Boholm puts it:

> In the modern discourse of risk, knowledge of contingency is delegated to the realm of expertise. When deploying the concept of risk, an expert approaches uncertainty as a dimension that can be analytically separated from its immediate context. An expert on risk analysis today is a scientific specialist and not a merchant, gambler or military campaigner, and such specialised expertise provides a privileged knowledge position in modern society. (2015, p. 6)

In biomedicine, the emergence of experts of all kinds is perhaps especially notable. Describing the novel form of biopolitics that operates in the twenty-first century, Nikolas Rose (2007) claims that new figures of authority and expertise have emerged in recent decades as "pastoral experts" (p. 6), meaning the growing range of professionals needed to face the personal, medical, and ethical dilemmas derived from the new technoscientific possibilities. Experts are therefore needed to navigate the decisions of the new "biological citizens" (Rose & Novas, 2005) who are expected to take responsibility for their own health and futures. As Sheila Jasanoff (1999) points out, paraphrasing Winner (1986), in this dialectic, experts have a privileged position: "Risk-talk implicitly empowers some people as experts and excludes others as inarticulate, irrelevant or incompetent" (Jasanoff, 1999, p. 137). In this regard, the risk perception of young women is influenced by the role of the biomedical figures of authority they encounter in the clinic. As Andrea noted, how could an 18-year-old donor question a doctor's authority?

In this sense, the first thing to consider when thinking about risk perception around egg donation is that, despite the level of information provided or neglected in fertility clinics, egg donation is a legal practice in Spain. This fact alone has to be regarded as an important aspect in building the trust of women interested in enrolling in an egg donation program. In a field involving high tech procedures and biomedical expert knowledge intermediaries, legality in a welfare state such as Spain is perceived by some as inherently warranting trust. In contrast, some donors were surprised and worried when they found out—usually during an interview with me—that egg donation is not legal in all European countries. The legal framework had an important role in the trust they had placed in the clinics and doctors. This was especially true among donors whose knowledge about the egg donation bioeconomy was delimited by the information provided by the clinics; others, instead, had started to challenge the official assumptions before our interview.

Andrea highlighted the relevance of the age at which decision-making takes place. This aspect was raised by other participants as well, especially by those who had donated years earlier. For most of them, their personal situation had changed significantly: Now they enjoyed more stable working conditions, higher education levels, and in some cases, had formed families of their own. In the interviews, most of the women who had donated years earlier claimed to have acquired a more critical perspective over time and questioned their ability to understand and consent at a much younger age, when they underwent the cycles. Biographical temporality emerges here again as a core category for understanding egg donors' rationalities and possibilities of risk anticipation and management.

Tempting Luck: Balancing Speculative Future Risks and Present Needs

Alba was one of the participants who highlighted her inability to fully understand and make an informed decision when she donated at age 19. She said:

> So when you're that age and they tell you that, for doing nothing, they will give you 1000 euros and that moreover you are protected and so on… and if you're thinking that you'll make someone happy, it's superb, isn't it? So, you do it. I don't know, they recruit you, it's not that you do it because you want it. They end up catching you, and of course there will be so many people that went because, "My friend did it, so I'm doing it as well." Try to imagine how many *niñas* [little girls] did it like this. […] Yes, they gave me

the information and made me sign that I had understood. Obviously, I hadn't understood anything. [...] What I was supposed to understand, I couldn't understand it at that time.
(Alba, 25 years old, one-time egg donor)

My interview with Alba occurred six years after her donation, and she told me that in recent years she had been thinking more about the donation. She confessed that the donation was something that she "couldn't quite reconcile 100%" from a feminist point of view. Similarly, my interview with Sonia took place 10 years after she donated at the age of 21. At that moment, her mother had just lost her job. Sonia didn't want to put more financial pressure upon the family and decided she would find the way to pay for her studies by herself. She did the donation in secret:

It's aggressive. It's an aggressive method and this... They don't tell you that at the clinic. They didn't give me any information obviously. Well, I expected that already. Well no, I didn't expect it, I mean, I couldn't imagine it. You don't imagine anything. You trust people, don't you? Because they're doctors, they're experts or whatever, and you don't imagine that this might cause you... that this can have side effects. Even less at 21. Now I might have a more critical view about the world, and I would think twice. At that moment, I didn't... I had faith that nothing would happen.
(Sonia, 31 years old, one-time egg donor)

What my participants' accounts share is the relevance of temporality in egg donation. First, some participants described their ability to assess risk at a young age as limited, and they pointed to the need for a certain level of "maturity" and "education" to be able to make this decision responsibly. Second, it was precisely as young women—age 18 to 35—that they were eligible to donate eggs in Spain (although most clinics had a cut-off age of 29 to assure optimized cycles). Third, this stage of life was often when women felt the most financially vulnerable. While most of them were not fully financially independent yet, they had expenses and responsibilities for which they could not or did not want to depend on others.

A final crucial aspect to consider related to temporality is that risk is a way to quantify and manage uncertainty (Boholm, 2015; Caplan, 2000;

Jasanoff, 1999; Lupton, 2013; Rosa, 1998). In other words, risk is itself a temporal concept, an assessment for a better future. Thinking about risk means thinking about the future. The notions of risk and risk perception give rise to several questions: Who is in the position to make choices directed to prevent worse futures? Who can invest in the future despite the present?

Coming back to the case of Mica, despite the new information I gave her about the possibility of unknown long-term effects, in her experience with egg donation, the pain lasted for *only* one week. She had a lot of expenses, and it is not hard to imagine how stressed she might have been starting the second year of a full-time academic program and simultaneously working two jobs. From her point of view, while the future risks were unstudied and therefore speculative, there was no doubt that she needed money to meet the expenses of the beginning of the school year. Her choice was directed to the present because the urgency she was facing was also situated in the present or immediate future.

The case of Mica resonated with that of another participant. Martina told me in our interview that she didn't want to pay much attention to the risks, precisely, because she *didn't want* to back out. She was counting on the compensation and, although she believed to some extent that the donation process could bring her negative consequences, she still chose to do it. She described her position of deliberate ignorance in this way:

> So, it's like… I feel bad saying this, but it's a fast way… but, of course, I'll say it again, I've never been aware, any of the four times, because I didn't look it up, I didn't really take interest in what I was doing. Well, I knew it because they had told me, but to what level? To what point? I always… I don't know, these last two times it was like, "Come on, I want money, I'll do it." Like very honestly, I mean… I sometimes felt bad for doing this to my body. I mean, it's not like I'm killing it… but maybe I am. I don't know.
> (Martina, 22 years old, four-time egg donor)

Just like Mica highlighted in her voice message, Martina pointed out the rapidity of the process, referring again to the urgency of the present despite the possible slow "killing" of her body she sometimes felt guilty about. Her choice was directed to the present because the urgency she was facing was also situated in the present or immediate future. This brings the individual and collective question of what futures matter, and toward whose bodies the affective condition of anticipation is projected (Adams et al., 2009). Who is called to enact the future in the present in an act of self-care and who is called to suspend concerns about the long-term future in order to have a slightly better present and immediate future? The balance between the unknown but somehow imagined future risks and the financial outcomes desired for the present was nicely articulated by Lucía:

> **Anna:** Do you think you will go through more cycles?
> **Lucía:** No. No. No, because I think that four times is… quite a lot. I mean, for health reasons I feel that… I've already tempted luck too many times.
> (Lucía, 22 years old, four-time egg donor)

The notion of "tempting luck" is interesting, because it offers a clear image of how some egg donors assessed and managed risk. Egg donation was perceived as an option that entailed risk, and every cycle was perceived as playing another round of roulette. Experiencing negative effects was seen as a matter of (bad) luck, in which the chances for complications increased with every spin of the wheel. On the other hand, egg donation was a reliable source of money. The amount of money they would receive and the time invested were optimized if they donated more than once (as they received a slight increase in the economic compensation for each cycle, and they did not have to undergo some of the medical checks and tests in subsequent donation cycles). They assessed the number of times they would spin the wheel against other variables such as their level of need and the pressure they were experiencing. Mica thought she had enough with one cycle; however, when facing financial stress later, she decided to take the risk of a second. Lucía, on the other hand, felt she needed to stop, as her lucky streak might end after four "successful" goes. However, as not many participants knew, the ultimate health consequences of egg donation may take years to be revealed.

Conclusions

In this chapter, I have discussed the ways in which structural factors and clinical practices in Spain circumscribed egg donors to a limited temporality, in which compressed information, pressure to sign informed consent documents quickly, the anonymity framework of egg donation, the young age of donors, and financial insecurity intersect, ultimately taming women's imaginaries and encouraging them to become egg donors.

First, I have demonstrated how the framework of anonymity in egg donation, together with the lack of information regarding the long-term risks involved, presented egg donation as an act without future consequences for donors. This erasure extended not only to the identities of recipients and offspring, but also women were denied any information on the future distribution and use of their eggs.

Secondly, I have explored how donors were situated within the present moment and action through a series of practices concerning how information about egg donation was provided and how intending egg donors gave consent. Within the context of informed consent processes at clinics, I have highlighted the practice of giving women all the informational documents and consent forms to sign the first day that they came to the clinic.

Finally, I have examined how egg donors were also circumscribed in time in relation to their young age and their position of financial insecurity. Here I have drawn on the experiences of several participants who donated years ago who highlighted issues such as their trust in institutions and doctors, their cultural capital at that time, and how their young age had an important role in their capacity for anticipating and managing risk. I have pointed out that risk anticipation is related to an ability to think about and invest in the future—an ability that was restricted by the circumstances of many of the donors in this research.

I argue that these arrangements result in a limitation of the participation and knowledge of egg donors and their separation from their product value, which turned donated eggs into bio-objects owned by the clinic whose futures would depend on market dynamics. The information arrangements can be seen as ultimately precluding any potential debate or claims for redistributive justice and asserting the clinic's ownership of any eggs extracted.

This analysis contributes to the growing interest in temporality and anticipation within new reproductive technologies by portraying how they rely upon temporal choreographies: While some subjects were called to anticipate their future through consumer choices directed to control and manage infertility, other subjects were led to suspend any discussion of risks to their future so that they could feed the demands of the growing fertility industry.

In the following chapter, I continue the discussion by focusing on the "operations" required by intending egg donors to make themselves "bioavailable" (Cohen, 2005). I contend that the labor of donorship goes beyond taking medication and undergoing extraction surgery to include choosing a clinic, learning about one's biocapital in the reproductive industry, arranging who will administer the injections and when, managing pain and discomfort, and juggling donation with other commitments, a process that subjectivizes and tames them in different ways.

References

Adams, V., Murphy, M., & Clarke, A. (2009). Anticipation: Technoscience, life, affect, temporality. *Subjectivity, 28*(1), 246–265. https://doi.org/10.1057/sub.2009.18

Beeson, D., & Lippman, A. (2006). Egg harvesting for stem cell research: Medical risks and ethical problems. *Reproductive Biomedicine Online, 13*(4), 573–579. https://doi.org/10.1016/S1472-6483(10)60647-5

Beskow, L. M., Hammack-Aviran, C. M., & Brelsford, K. M. (2020). Developing model biobanking consent language: What matters to prospective participants?(Report). *BMC Medical Research Methodology, 20*(1), 1–14. https://doi.org/10.1186/s12874-020-01001-2

Beskow, L. M., & Weinfurt, K. P. (2019). Exploring understanding of "Understanding": The paradigm case of biobank consent comprehension. *The American Journal of Bioethics, 19*(5), 6–18. https://doi.org/10.1080/15265161.2019.1587031

Bierschenk, T. (2019). Postface: Anthropology, bureaucracy and paperwork. *Journal of Legal Anthropology, 3*(2), 111–119. https://doi.org/10.3167/jla.2019.030207

Boholm, A. (2015). *Anthropology and risk*. Routledge.

Borrelli, L. M., & Lindberg, A. (2019). Paperwork performances: Legitimating state violence in the Swedish deportation regime. *Journal of Legal Anthropology, 3*(2), 50. https://doi.org/10.3167/jla.2019.030204

Brody, J. E. (2017, July 10). Do egg donors face long-term risks? *The New York Times*. https://www.nytimes.com/2017/07/10/well/live/are-there-long-term-risks-to-egg-donors.html

Caplan, P. (2000). Introduction. In P. Caplan (Ed.), *Risk revisited* (pp. 1–28). Pluto Press.

Cattapan, A. R. (2016). Good eggs? Evaluating consent forms for egg donation. *Journal of Medical Ethics, 42*(7), 455–459. https://doi.org/10.1136/medethics-2015-102964

Coeytaux, F., & Tober, D. (2013). *Eggs for cash: Pitting choice against risk*. Rewire. Retrieved February 26, 2018, from https://rewire.news/article/2013/09/04/eggs-for-cash-pitting-choice-against-risk/

Cohen, L. (2005). Operability, bioavailability, and exception. In A. Ong & S. J. Collier (Eds.), *Global assemblages: Technology, politics, and ethics as anthropological problems* (pp. 79–90). Blackwell Publishing.

Jasanoff, S. (1999). The songlines of risk. *Environmental Values, 8*(2), 135–152. https://doi.org/10.3197/096327199129341761

Lupton, D. (2013). *Risk*. Routledge.

Molas, A., & Whittaker, A. (2021). Beyond the making of altruism: Branding and identity in egg donation websites in Spain. *BioSocieties, 17*(2), 320–346. https://doi.org/10.1057/s41292-020-00218-0

Molas, A., & Whittaker, A. (2023). Tempting luck: Temporalities and risk anticipation among egg donors in Spain. *Science, Technology, & Human Values, 49*(6), 1256–1280. https://doi.org/10.1177/01622439231164666

Namberger, V. (2019). *The reproductive body at work: The South African bioeconomy of egg donation*. Routledge.

Niemiec, E., & Howard, H. C. (2020). Germline genome editing research: What are gamete donors (not) informed about in consent forms? *The CRISPR Journal, 3*(1), 52–63. https://doi.org/10.1089/crispr.2019.0043

Novas, C. (2006). The political economy of hope: Patients' Organizations, Science and Biovalue. *BioSocieties, 1*(3), 289–305. https://doi.org/10.1017/s1745855206003024

Pearson, H. (2006). Health effects of egg donation may take decades to emerge. *Nature, 442*, 607. https://doi.org/10.1038/442607a

Rivas, A. M., Lores, F., & Jociles, M. (2019). El anonimato y el altruismo en la donación de gametos: la producción de biocapital en la industria reproductiva. *Política y sociedad, 56*(3), 623–644. https://doi.org/10.5209/poso.60564

Rosa, E. A. (1998). Metatheoretical foundations for post-normal risk. *Journal of Risk Research, 1*(1), 15–44. https://doi.org/10.1080/136698798377303

Rose, N. S. (2007). *The politics of life itself: Biomedicine, power and subjectivity in the twenty-first century*. Princeton University Press.

Rose, N. S., & Novas, C. (2005). Biological citizenship. In A. Ong & S. J. Collier (Eds.), *Global assemblages: Technology, politics, and ethics as anthropological problems*. MBlackwell Publishing.

Schneider, J. (2008). Fatal colon cancer in a young egg donor: A physician mother's call for follow-up and research on the long-term risks of ovarian stimulation. *Fertility and Sterility, 90*(5), 1–5. https://doi.org/10.1016/j.fertnstert.2007.12.074

Schneider, J., Lahl, J., & Kramer, W. (2017). Long-term breast cancer risk following ovarian stimulation in young egg donors: A call for follow-up, research and informed consent. *Reproductive Biomedicine Online, 34*(5), 480–485. https://doi.org/10.1016/j.rbmo.2017.02.003

Thompson, C. (2005). *Making parents: the ontological choreography of reproductive technologies*. MIT Press.

Tober, D., Garibaldi, C., Blair, A., & Baltzell, K. (2021). Alignment between expectations and experiences of egg donors: What does it mean to be informed? *Reproductive Biomedicine & Society Online, 12*, 1–13. https://doi.org/10.1016/j.rbms.2020.08.003

Waldby, C., & Mitchell, R. (2006). *Tissue economies: Blood, organs, and cell lines in late capitalism*. Duke University Press.

Winner, L. (1986). *The whale and the reactor: A search for limits in an age of high technology*. University of Chicago Press.

Woodriff, M., Sauer, M. V., & Klitzman, R. (2014). Advocating for longitudinal follow-up of the health and welfare of egg donors. *Fertility and Sterility, 102*(3), 662–666. https://doi.org/10.1016/j.fertnstert.2014.05.037

Woodruff, E. (2017). 'We simply don't know': Egg donors face uncertain long-term risks. *Statnews*. https://www.statnews.com/2017/01/28/egg-donors-risks/

5

Laboring Donors

Introduction

Mica lived in a town 27 kilometers from the clinic where she was undergoing her first egg donation cycle in exchange for a very much wanted 1000 euros cash (see Chap. 3). To get to the luxurious private clinic where she was undergoing the process, she had to take the train, the light rail, and a bus, and walk uphill for 10 minutes—a trip that took at least 1 hour and 15 minutes. After the appointment at the clinic, however, she also had to go to school, which took her another hour of public transport. Then, at 9 pm, she took another hour-long train ride home. Mica repeated this ritual every day: she had to go to the clinic to undergo tests or ultrasound monitoring or to pick up medication. Toward the end of the process, however, something disrupted her careful scheduling.

> **Mica:** One day, the train was delayed, and I had to get home. I had gone to the clinic, I had all the needles with me, and I had to prepare the whole *show* [referring to the preparation of the injections]. So, I got off at a suburb some station before my own stop, because I had to do it at 10 pm, I had to be very punctual. And I couldn't jab myself. My body couldn't… my hand stopped every time. I was going to jab myself and at the last moment…

A. Molas, *Taming Egg Donors*, Health, Technology and Society,
https://doi.org/10.1007/978-981-96-5103-0_5

> my hand stopped. It was like my head going "jab," but my body didn't want to do it anymore. It was the last one [the last injection of the medication process], I was very hormonal. I think the next day, or the following, I was having the extraction, and this is what I remember the most… I don't know if I would do it again really because my body was rejecting it, I mean, my hand wouldn't go all the way down. My hand stopped and I tried again and again but at the last moment… and then I jabbed myself but of course the needle couldn't really go in [because her hand was stopping short before completely puncturing her skin] and I started to feel bad, because this was a very, very specific injection. For the rest of them, if you do it 5 minutes late it's not a big deal, but this one… and I remember getting home and saying [she exhales], "My body doesn't want this, and I don't want…" I mean, everything's alright until my body says it's enough. […]
>
> **Anna:** But did you finally manage?
>
> **Mica:** Yes, in the end… [I was able to jab myself]. [It was hard] because I was in Sant Adrià, which is an area that at 10 pm… and I had a lot of needles! […] And there was a group of guys around and I started to cry… because of course I was very hormonal with all of this. […] I was okay when I got off the train, but when I saw that I had to jab myself, but my hand wouldn't go down… and I tried three times, there I started the get anxious, "No, Mica, enough."

Mica presents herself as being in a struggle with her own body, which she describes as a separate entity that spoke to her desperately. As her story shows, Mica's daily commitments were difficult to balance with the requirements of egg donation. Juggling work, studies, and the medication was stressful and challenging. During my fieldwork, I encountered many stories like Mica's, which portray the amount of labor egg donors performed during the cycle, labor that goes far beyond the stimulation process and the extraction surgery. In this chapter I unpack these different kinds of labor, claiming that they are generally overlooked in the growing literature on reproductive bioeconomies.

Lawrence Cohen (2005) proposes the concepts of "bioavailability" and "operability" in his examination of kidney donation in India. In his study, he explores the social apparatus that governs kidney redistribution in South India. According to him, to be bioavailable "is to be available for the selective disaggregation of one's cells or tissues and their reincorporation into another body (or machine)" (p. 83). The bioavailability of large populations is necessary for the creation of global forms of organ and tissue circulation across different expanding global biomarkets. However, the process of bioavailability does not come naturally. As he argues, for subjects to become bioavailable, they need to be situated within a specific technoscientific regime allowing and underpinning the transformation of some subjects (usually the poor and marginal). These subjects, in turn, have to embrace these transformations often framed in progressive discourses embedded in neoliberal ideas of entrepreneurship, care, and decision-making—a process that Cohen (2005) names "operability." As the author puts it, operability "is thus necessary to remake one's mindful body in accordance with the demands of developmental modernity, to remake one *as if* one were modern. If the operation becomes a form through which constitutively marginal, pre-modern subjects can secure some form of modern participation in the nation-state, it may become a critical desideratum" (Cohen, 2005, p. 87).

Different accounts in the fields of reproduction and science and technology studies have captured the complexity of the processes that govern the production of subjectivities and value in reproductive bioeconomies (e.g., Nahman, 2013; Pavone & Goven, 2017; Rudrappa, 2015; Siegl, 2023; Weis, 2021). Additionally, post-Fordism has brought the externalization of social welfare and risks to employees, giving rise to a new subjectivity shaped by the corporatization of states and industries. The new subject is at once "in charge of itself, responsible for itself, yet an instrumentalizable and potentially dispensable element of the whole" (Brown, 2015, p. 38).

The frame of bioavailability has been used in the examination of the process through which women make themselves ready for the global reproductive market in different ways. For example, Laura Perler and Carolin Schurr (2020) use it to argue that the assemblage that makes women in Mexico bioavailable is their "reproductive biographies," for which they claim that "[women's] affective experiences of (unwanted) pregnancies, abortions, family planning, and sexual, physical and obstetric violence are closely entangled with their motivations to make their oocytes bioavailable" (p. 6). In contrast, Diane Tober and Charlotte Kroløkke (2021) use the concept of bioavailability to highlight the need to infuse the narrative of the "epic quest" of all participants in the fertility industry into each step of the process (from egg donor recruitment to oocyte transactions), in order to sustain a bioavailable workforce—a process that they label as "emotional choreography."

These accounts show how clinical labor under neoliberalism builds particular kinds of subjectivities. Less research has been undertaken to define the types of embodied labor that egg donation demands. In her Marxist/feminist analysis of the egg donation bioeconomy in South Africa, Verena Namberger (2019) argues that the "fetishization" of oocytes as key bio-objects for the generation of value in reproductive bioeconomies obscures the role of egg providers in their production, for which she claims that "[t]he (re)productive and working body in the bioeconomy is a weakly illuminated, if not completely blind, spot" (p. 8). Building upon her observation, in this chapter I reveal the kind of labor that egg donation entails, the conditions in which it is undertaken, and how the process produces specific subjects in the reproductive bioeconomy.

The concepts of "bioavailability" and "operability," then, are used in the context of this book to shed light on the steps ("operations") that women have to undertake to make themselves "bioavailable" for the global reproductive industry. This analysis of operability as a process through which individuals from marginal backgrounds can actively engage with modern society informs this research, in which women find in egg donation a way to secure themselves a future—or a better present—in an industrialized post-Fordist context. Connecting Cohen's approach to the link between women's decision to become egg donors and their position in the labor market explored in Chap. 3, I conceptualize these operations not so much as steps for the clinical labor to begin, but as forms of labor that themselves produce value in the reproductive bioeconomy. In other words, I contend that becoming an egg donor in Spain is a laborious ongoing process—both physical and emotional—that takes place before, during, and after the treatment itself. I also argue that it brings with it a process of subjectification and taming that facilitates egg donors' alignment with clinics' extraction purposes.

In describing these operations, I first discuss how intending egg donors chose a clinic to undergo their cycle, showing how the perceptions of prestige and luxury associated with private fertility clinics emerged as indicators of safety and care. Second, I address the realization of intending egg donors that they might not be able to choose the clinic at which they would undergo the procedure. By focusing on the experiences of participants who were rejected from a donation program, I highlight how the clinics' selection criteria made women think of themselves in terms of biocapital for the reproductive bioeconomy. Third, I discuss the self-medication regimes and work arrangements egg donors arranged during the cycle, pointing out how they were rarely directed to increase their own comfort during the process but rather to juggle their jobs and other commitments with the donation. Finally, I explore how egg donors managed the pain and discomfort experienced during the process, challenging fertility clinics' dismissal of the pain as "similar to menstruation, but stronger."

Choosing a Clinic: Luxury and Safety in the Private Sector

> I went there because it was like "Institut Marques," "Pedralbes," I mean… there are a lot of things with that, right? You need to be careful; it's also something that affects your body a lot. It affected me a lot at the hormonal level, then I also found out that they did it through Sanitas[1] and so on, and that also helped me… I really liked that, right?
> (Andrea, 29 years old, two-time egg donor)

Doing fieldwork for my research involved walking around a particular part of Barcelona very often. The area from Francesc Macià Square to Maria Cristina Square and connected by Diagonal Avenue is home of the majority of fertility clinics in Barcelona, as well as the headquarters of big corporations, banks, private hospitals, and hotels. Surrounded by the four wealthiest neighborhoods of the city (Les Corts, Sant Gervasi, Sarrià, and Pedralbes), this part of Diagonal is like a noisy valley surrounded by high buildings, new and old, which act as borders for one of the fanciest parts of the city (Illustration 5.1). Women intending to donate their eggs

[1] Private health insurance.

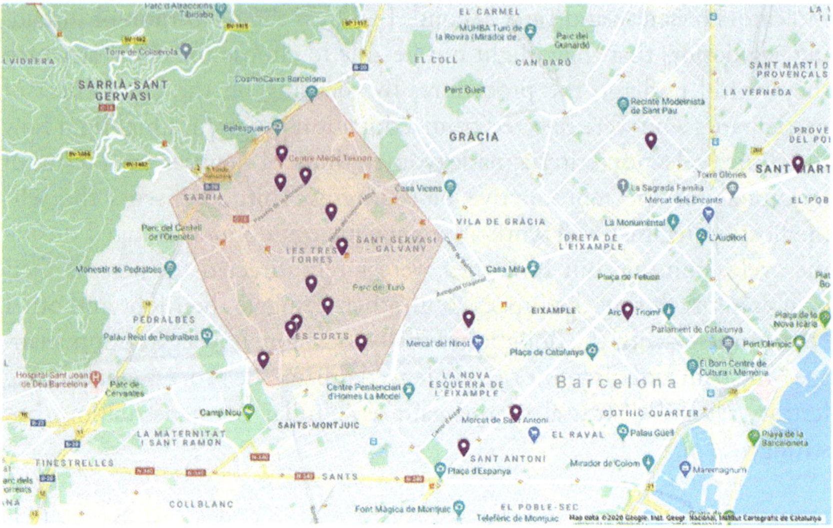

Illustration 5.1 Map of Barcelona fertility clinics

usually had to stroll around these streets for some time. The ascent, both real and symbolic, to this part of the city shaped egg donors' experiences in different ways. For most of them, who did not live nearby and sometimes not even in the city, this would become a pleasant liminal space which would allow them to briefly enjoy the delights of a luxurious industry that would invest in their care for a while.

The region of Catalonia has more than 38 fertility clinics, most of them in the city of Barcelona (SEF, 2022). Choosing a clinic to undertake egg donation is one of the first forms of labor that donors must perform. Interestingly, even donors who showed complete alignment with the clinics' discourse of good faith and claimed altruistic motives struggled to choose a clinic. The impression that they needed to "be careful" about where to make their donation, as Andrea said, reveals the perceived shadows around the industry.

Both donors and clinicians joked about the donors' concerns that doctors might take their kidney as well. Indeed, the imaginary that some clinics did not perform ethically was shared by most of the participants and was the reason why most women took significant steps and time to select a clinic perceived to be "safe." The most important factors were money, location, and safety. But as the amount of money paid under the term "economic compensation" does not vary much from clinic to clinic, and as most of Barcelona's clinics are in the same area, safety became the most important consideration. Consequently, personally knowing someone who had had a good experience at a clinic was decisive.

> Well, it was… I looked it up on the Internet and there were a lot of clinics, but of course, I wasn't going to… It was something that concerned me a bit, right? Then finally I talked to the friend who told me about it for the first time and I asked her about the clinic where her sister went.
> (Mireia, 28 years old, one-time egg donor)

Interestingly, ideas around safety were usually intertwined with perceptions of the luxury and prestige of the clinics. The fact that the egg donation sector is privatized in Spain meant that participants often associated quality and safety with the brands of private clinics, with many donors wanting to undergo the cycle in clinics that "had a name."

> I remember very well looking a lot on the Internet, right? I checked the places where I could do it, what clinics did it, which one was close to the university… and which one was paying the most fundamentally, because there are some that pay more and others that pay less. Then I finally chose CEFER clinic. In fact, I visited a lot of websites and they started to call me from different places, because when you start looking, they already start to call you and…[2] Finally, I was keen on CEFER clinic. I don't know exactly why. I think it was because I got the impression that, being the Instituto Teknon[3]… right? So, it gave me the feeling that it was better in the sense that it was not a small clinic. Because there are some clinics that are smaller, right? So, I don't know, that gave me a kind of security.
> (Lucía, 22 years old, four-time egg donor)

[2] Most egg donation clinics' websites have a form on the homepage for intending egg donors to leave their contact details so that the clinic can contact them.

[3] Private hospital in Barcelona since 1994 known for its luxurious installations, advances in medical technology, and famous patients.

> So, of course, I searched, and I saw Dexeus, and I said, "Wow, that's the most renowned one." I was [deciding] between Dexeus and Teknon, and I decided Dexeus because there were always ads on TV and things like... that focused on motherhood issues, for women. I don't know, it's the impression that I had. Most football players or famous people in Barcelona give birth there, right?
>
> (Helena, 27 years old, two-time egg donor)

Similarly, Angela (22 years old, two-time egg donor) highlighted that the clinic where she made her first donation was the one "where the king of Spain was going."

Indeed, regardless of their main motivation to make the donation, egg donors reported making the decision consciously to avoid being taken advantage of. These concerns can be read as a means to have a degree of control over a process that generates fear and a sense of vulnerability (Warren et al., 2013).

Donors enjoyed spending time in this fancy part of the city and receiving the attention of clinic staff. Participants talked about the beautiful settings of the clinics and the comfort they always found there, highlighting details such as having free food and coffee, which helped them feel reassured about their decision.

"Benchmark professionals": Perceptions of Reproductive Health Care Services

In addition to the prestige of the clinics and their settings, donors were also attracted to the prestige of the doctors in assisted reproductive medicine. Maria, a 34-year-old midwife, had an unwanted pregnancy a month after the donation, which was in part due to the hormonal alteration from the IVF medication. Expressing guilt and anger, she told me about her extreme distress when she got the news:

> *Hostia puta* [fucking hell], I mean, I had already had a period after that [the extraction], so I assumed that that extra fertility was over, but in reality, it put my cycle out of order in a way that I ovulated when I didn't expect to, and I got pregnant. And that was like, "fuck it", right? I mean, the bitter

part of all of this, because I… of course it was my responsibility, but it was a bit because of this hormonal imbalance due to the process.

Maria decided to terminate the pregnancy. Although she worked as a midwife in the public health system, she chose a private clinic instead of a public hospital. When I asked about this decision, she explained the risk to her professional reputation. If the abortion were performed in the public system, it would appear in her electronic medical history and would be visible to health providers who cared for her in the future:

Hostia [Damn], I mean, I'm a midwife; you have all of this under control. This, this stain can't be on your file. I didn't want this chapter of my life in my medical history, right? A voluntary abortion because you haven't looked after your fertility as you should have, right? You spend the day advising people not to have unprotected sex and now it's you who have it and so on, right?

She decided, therefore, to undergo the termination in a private clinic for a price of around 400 euros, "So finally part of the money from the egg donation went to pay for my abortion," she said with irony. Maria stressed the differences she perceived between the settings and professionals of the fertility clinic and the abortion clinic.

If we compare it with abortion clinics, come on, I have the impression that they also move a lot of money and still they have very seedy clinics. In contrast, here [in the fertility clinic] the facilities, the resources… I don't mean that the chairs are cool, but the ultrasound scanner is very good too. Or the professionals, right? The prestige they have, their careers. I can't say everyone, but in many assisted reproduction clinics, the professional trajectory of their professionals has nothing to do with the ones who end up in pregnancy termination. […] There are no benchmark professionals that work in pregnancy termination. […] The social status of the gynecologists that are in assisted reproduction has nothing to do with the ones in pregnancy termination. Usually, they end up being gynecologists who don't have their title validated and that come from South America and end up [working] at abortion clinics, you know? Instead, the ones in assisted reproduction are another vibe. They're professionals that work in the

> private [sector] but also in the public [sector] and end up being benchmark professionals. They are professors at the university… I mean, it's totally another kind of professional. […] They're both moving a lot of money a priori if you think about it from the outside, right? But one seems that it's investing the money and in the other [abortion clinics] probably there's a *capo* [mafia boss] behind it all that is making a profit, but the professionals are not getting much out of it, no, no. Because I have no doubt that someone is making a profit from the money from abortions. There is a lot of money moving… where is it?
>
> (Maria, 34 years old, one-time egg donor)

Her experience provides an interesting example of the differences between the reproductive health systems available for women in the private sector, as well as the perceptions around them. While she experienced care, luxury, and a perceived high level of expertise when undergoing donorship, she encountered a seedy operation when she needed a reproductive treatment (an abortion) for herself. If private fertility clinics direct money, time, and reproductive and emotional labor to achieve pregnancy (seen as a major accomplishment), private abortions clinics offer shabby facilities and lower-profile professionals to terminate pregnancies (in this context, seen as a source of shame). What is extremely valuable in one context becomes waste in the other one (Illustration 5.2).

ABORTION CLINICS	FERTILITY CLINICS
Low-profile professionals	High-profile professionals
Unwanted pregnancies	Highly desired pregnancies
Seedy clinics	Luxurious clinics
Morally problematic	Morally good
Dubious financial structure	Wealth distributed and invested in the care of the users

Illustration 5.2 Maria's perception of the different reproductive services in place in the private health sector

We see here how women's reproductive capabilities and bodily fluids have vastly different social value in one setting compared with the other. As Charlotte Kroløkke (2018) puts it, their value is "intertwined with the cultural values in which reproductive parts and social relations are embedded" (p. 3). The social relations embedded in the Spanish private sector of abortion and assisted reproduction become clear in Maria's story. The services related to the termination of pregnancy are regarded as morally ambiguous both because of the practices underlying them and because of the kind of professionals working in there. In contrast, assisted reproduction services seem to buzz with the same hope and excitement that emanate from their high-tech and scientifically sophisticated equipment.

Abortion's moral troubles are traditionally understood to lie in the destruction of human life. This feature is in fact shared with assisted reproduction, in which extra embryos are produced and later destroyed. I argue that fertility clinics animate imaginaries that point to the life they will create rather than the life that will be destroyed in the process. Ideas of family, motherhood, and the reproduction of a social class are coordinated in ways that make it possible to perceive the clinicians of these industries as "benchmark professionals."

Understanding One's Biocapital

If, as previously mentioned, the process of selecting a clinic was the first stage in the making of a donor, donors' realization that they might actually not be able to choose was the second. Most potential egg donors came to understand this when they started receiving calls from the clinics, which happened as soon as they filled out the clinics' online contact forms. What usually started as an assessment to determine which clinics offered the best combination of money, location, and safety sometimes ended in a quest to find a clinic that would accept them at all. I describe the rejection of donors in detail in Chaps. 6 and 7. The main reasons are having certain phenotypical features "in surplus" or not meeting the basic health, genetic, and psychosocial quality standards of the clinic.

The selection criteria of fertility clinics raised issues that participants had previously not perceived to be a problem. In this sense, the criteria constitute an avenue for thinking about their bodies and their health in a new way: in terms of biocapital, desirability, and future reproductive potentialities. When answering the questions of clinicians and undergoing the multiple required screenings, egg donors started to grasp the biocapital they possessed in the clinic, and then they tried to find means to convert it.

This process became evident when I observed how some women did not simply accept a clinic's rejection. Instead, many participants called other clinics if they had been rejected by the first one they had applied to. Some did it when they considered the clinic's reason for rejection—such as height or weight—to be baseless. Others, however, turned to a new clinic after being rejected by the first one, this time withholding the information that had led to the rejection, for example aspects of their own or their families' health histories.

> **Mica:** And I called, and they rejected me because of my height, I remember. […] Because I'm 1.56 [meters] and they said "no," that I was very small, and thanks very much [for calling]. And I said, "What?"
>
> **Anna:** How did the conversation go?
>
> **Mica:** No, I mean, it was straightforward, they said "Hi, how are you?" and I, "Good, I'm interested, whatever" and they, "And what do you look like?" and I, "I'm blond, normal constitution, green eyes…", "How tall are you? How much do you weigh?" […] "Oh, okay thanks, forget it." I mean everything was very polite but basically, they hung up. And I thought, "But there are many women who are…" I mean, my height is nothing crazy, my boyfriend is 1.40 [meters]. I mean… and that's it. They rejected me like that, very fast. After that, I talked to a friend… […] and she said, "I don't think they will reject you at the clinic where I did it." And there, they didn't have any problem with it. They said that it was okay so.
>
> (Mica, 22 years old, two time egg donor)

In a similar manner, Berta, a 21-year-old, two-time egg donor, opted to have her first visit at a private clinic associated with a famous private hospital in Barcelona. After filling in both the clinic's questionnaire and the psychological test, she was informed on the phone that she could not be a donor due to her mental health issues. By this, the clinicians meant Berta's aunt's diagnosis of a borderline personality disorder. Angry and frustrated by the outcome, Berta decided to continue her search. Apparently, the second clinic she went to did not check family mental health history so exhaustively, and she was accepted. Emily had a similar experience:

> **Emily:** Well, I looked for information, I started looking for clinics… for a week or so. I started to read what the process was about, how long it lasted… and then the demands of each clinic, because… that was another topic. Each clinic had different demands, at least the one where I was, I had the impression that they were not as strict in terms of… For example, they call you and one of the first things they ask you is your weight and height. Then the first clinic that called me they told me that I couldn't do it because my weight and my height made it complicated. They told me that in terms of the anesthesia it could be complicated.
>
> **Anna:** Too short?
>
> **Emily:** Too small, yes. And with a weight… well, that was at the limit of the body-mass index, so they said no. But then they called me from another clinic after a few days and at this one, they said yes. And they also asked me my weight and my height right away. I thought, "What's wrong here? Is it that at this clinic they're not strict?" Because at the end of the day, it's a medical and health matter, isn't it? Then presumably they should have more or less the same rules. If it's dangerous at one clinic it's also dangerous at another, isn't it? So, in the end, I still had this doubt.
>
> (Emily, 23 years old, two-time egg donor)

For some participants, the selection criteria of fertility clinics also implied a new understanding of themselves as racial beings. In this new definition, race did not only imply certain visible features. Even if not physically perceptible, within fertility clinics, race was understood as a potential, hidden in their genetic past, threatening the future of intending parents and children. In Chap. 7, I devote a section to the definition of race that was enacted at the clinics and how it impacted selection processes. In this section, instead, I focus not so much on how race is delimited but on how intending donors came to understand themselves through this lens.

Valentina, an intending egg donor from Colombia who had been living in Spain with her mother for 11 years, tried to donate eggs when she was 18 years old. As she told me, she was surprised about the interest of the clinic not just in her origins but also in those of her parents.

> **Valentina:** […] If I'm not wrong, they asked me about my family's origins. Where my family came from, where I had been born. […] But I remember it was our looks that interested them the most.
>
> **Anna:** Did you have to answer questions because your family was from Colombia?
>
> **Valentina:** Well, they asked about the origin of my mom and my dad, from where exactly…
>
> **Anna:** From where exactly in Colombia?
>
> **Valentina:** From where and what part of Colombia. I think they asked about the physical appearance of my parents as well. Mmh… I don't know if also about my sister. I remember that one time, the girl [nurse] was there and she commented that… I don't remember exactly how, but something like "Well, you don't really look like a Latin American." I'm white, I have fair hair… It was just a comment, I mean, that's it. I don't know if it was something good […], because of course, if I was *morenita* (brownish), small eyes, straight hair… I don't know if that, for example, is also an obstacle or if it's something that the clinic isn't interested in because the parents aren't interested… but yes, […] I think they commented that sometimes they asked about the looks of the parents, because many times a child doesn't look like the parents, but like the grandparents. Of course, if it's from my parents, especially my mom, she has some very characteristic features. I mean, they are more visible of… a person from Latin America.

Valentina came to understand that visible "Latin American features," as Núria, a (nurse) put it, were an obstacle in the reproductive industry, and further that this represented a risk that was inscribed in herself through the heritage of her parents. So, even if she did not display such features, as was the case, she was conceptualized as a potential carrier. After that phone call, she was still invited to have a first appointment at the clinic but was finally discarded anyway on the pretext of her height, which was 1.55 meters. The narrative around the rejection was not related to the risks associated with anesthesia, as for Emily, but on the differences between Colombian and Spanish features.

> She said it politely, she said "Listen, it's not that there's a problem. It's not that we aren't interested, but you need to understand that the process of extracting eggs is expensive," and basically, she kind of told me that it wasn't worth it to extract my eggs if the possibilities of someone choosing them to have a child were low. Because, of course, this means maintaining the eggs knowing that they won't choose me or that the probabilities are very low. So, it was when she said, "Well, it's not for any health related issue, it wouldn't have any bad effect on you, but the thing is that basically many times moms want a donor who's a bit taller." And I think she wanted to be conciliatory, "They want them [the child] to look like them physically, and physically the average height of moms in Spain is 1.60 [meters]." In Colombia, I'm average, I'm normal. There everybody is short, but here it's not like that.
>
> (Valentina, 21 years old, rejected from the egg donation program)

Through both the questions about her family's origins and features and the statements about Spanish women's average height, Valentina was constructed as risky and less desirable than Spanish donors, or than donors with a whiter and taller heritage in general. As she told me, after one year she was contacted again by the clinic saying they found a recipient "who didn't care, or who was also short, I don't even know." However, at that time she was anemic and, therefore, not medically able to undergo a cycle. Her racial biocapital at the fertility clinic, which she fully understood once she underwent the screening, would not let her choose when to do the donation, or even whether to do it at all.

A similar thing happened to Nancy, a participant who was born and raised in Peru until age 8 and then moved to Spain with her family. She was 22 when, following the example of two of her friends (Mica and Lucía, also participants in this research), she tried to become an egg donor. After reading numerous documents, signing informed consent forms, completing questionnaires about her personal traits and health history, undergoing a check on her weight and height, having a picture taken, and completing a psychological test (all at the first appointment), she was rejected because of her racialized features.

> **Nancy:** There was a very messed-up question that said, "What is your race?" Something like that. And I thought, "I don't know." [laughs] I mean,

my skin is brown but… I thought that this was very… unnecessary in part. Because of course I thought that this was more like an altruistic thing. You go there and you donate if you want to. Like when you donate blood. No one asks you, I don't know, if your blood is too brown or… very light red, right? And of course, they asked these things about my physical characteristics, so it was hard. I mean, it wasn't hard, but I thought it wasn't nice really. And they asked me like this, "What is your race?" […].

Anna: What options did you have?

Nancy: There were no options. I mean, you had to write what you were, and I thought, "What am I?" And of course, I asked her, and she told me, "Don't worry, we'll write it down ourselves." I guess that she had seen me and that she was going to write something down.

The day after she completed these forms, tests, and checks, she received a call from the clinic.

They called me and told me that it was better to cancel the appointment because there weren't people who wanted my profile… because my features… well, because of my ethnic features I guess, that it was better not to go through the whole process… because of course, these cells, my eggs, would stay there for months, and of course they didn't want to do it in case no one wanted those eggs and they were there with nothing to do, you know? They told me that if someone wanted my eggs, then they would call me to start the process, etc. Then the appointment ended up in nothing, and that's it. That was all I did.

Nancy's understanding after undergoing the screening was that the desirability of her phenotypical features was lower than the cost of the process itself. Or, in other words, the transformation of her bodily capabilities into marketable commodities was not considered economically worth the investment. When the nurse expressed doubt that anyone would want Nancy's eggs, she revealed the hidden logic that clinics used in selecting donors, a logic that became visible to intending donors during the selection process. It was through such conversations that egg donors came to understand themselves in terms of biocapital in the Spanish fertility industry.

Similar to invisible gendered reproductive labor in other spheres, the labor involved in selecting a clinic and filling out the multiple questionnaires and undergoing medical screenings was not compensated in any way. Fertility clinics, instead, focused on the cognitive labor of the scientists as the element that enabled them to assert property rights over living matter, which appeared to them as already available raw materials disentangled from the in vivo labor needed for its production (see Cooper & Waldby, 2014; Namberger, 2019).

Getting Organized to Take the Shots

Some participants reported finding it easy to get to the clinics, which were located a few hundred meters from the Zona Universitària, a 500 square-meter area that hosts more than 10 university schools and departments. However, not all donors were university students, and many of them had to plan their travel carefully so as to attend appointments at the clinic without breaking other commitments. Mica's story of giving herself a shot in the train station shows the level of coordination required to be able to attend the appointments, take the injections, and juggle the other commitments of daily life.

According to Mica (Images 5.1, 5.2, 5.3, and 5.4), when it became too much, her body itself put a stop to it. Mica described a strong dissociation from her body during the process. In her account, she presented her body as a separate entity that unexpectedly disobeyed her order to inject the medication. Similar to the experiences of the surrogates in Elly Teman's ethnography (2010) in Israel, for some donors, putting themselves at the disposition of fertility clinics meant suspending communication with their own bodies, which therefore needed to find different ways to express themselves. For Teman (2010), these bodily reactions "may thus be considered symbolic expressions of the lack of freedom and structural powerlessness under the restrictive contract and surveillance of the body politic, medical institution, and contracting people" (p. 45).

In this section, I consider the level of coordination and organization required to be able to attend appointments, take the medication, and

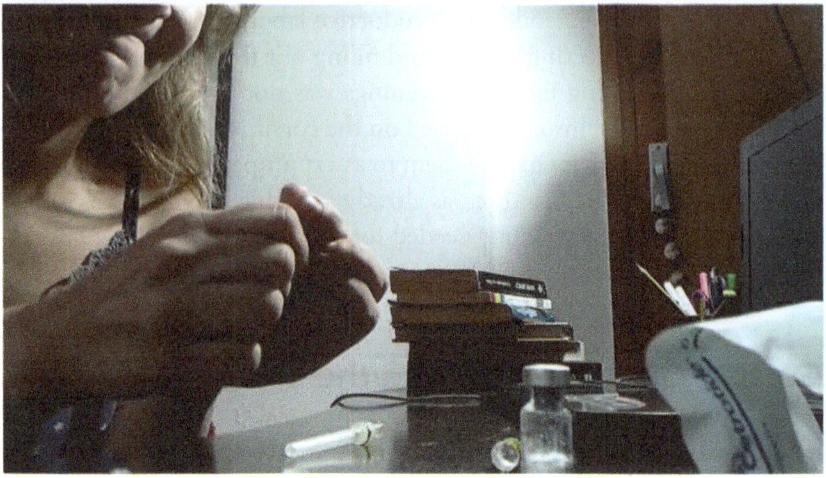

Image 5.1 Mica prepares the medication for the last injection of her second cycle (I)

Image 5.2 Mica prepares the medication for the last injection of her second cycle (II)

Image 5.3 Mica prepares the medication for the last injection of her second cycle (III)

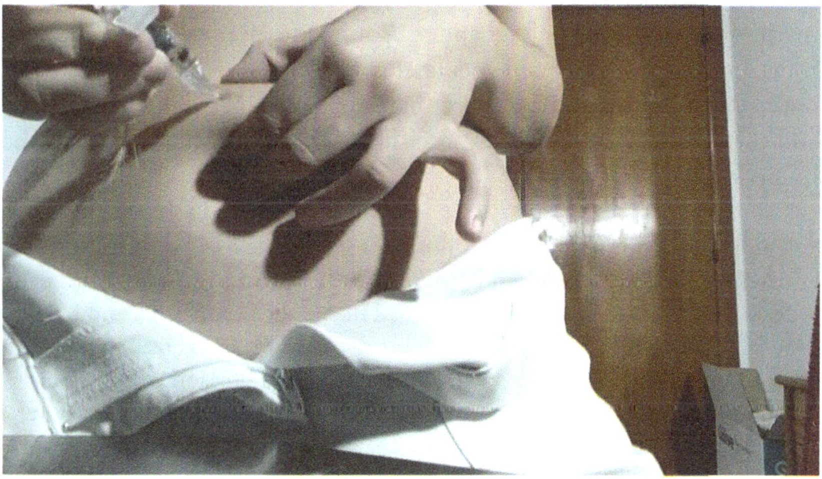

Image 5.4 Mica injects herself with the last dose of medication for her second cycle

keep other daily commitments. Mireia, a 28-year-old participant (22 when she underwent the cycle), struggled to balance the medication process with her studies and her work:

> It was crazy. I mean, I was up to my eyeballs all day, sincerely. Luckily… well I'm not sure if "luckily"… I was working, and I was going to classes very irregularly, you know? Then the truth is that I had to think very carefully about the schedule of the injections, right? Because I don't really remember how it was, but they always had to be at the same time. Of course, I had very unstable schedules, but then I realized that an hour before going to work was a good time. During that week, I checked my schedules very well and so on to try to do it always at the same time and to be able to be at home, because I had to keep the product to inject in the fridge. Luckily, it was just one week, but it drove me crazy to have to go to the clinic, to go home, to jab myself, to go to class…

As Mireia said, *when* and *where* to inject the medication were usually matters to be studied beforehand, but for some donors, especially first-timers, *who* would inject the medication for them also needed to be thoroughly planned. Martina, for example, relied on her flatmate to do it, and this caused her anxiety because she was afraid he might forget to be home for it.

> I didn't feel panic about needles because for example I had donated blood before, or tests and so on, but to jab myself…. The truth is that it made me feel… I hadn't thought about it till I had the needle in my hand, and I had to inject it. Until that moment I didn't say "shit," I hadn't realized that I had to do it. Luckily, I lived with a guy who was a physiotherapist, so he jabbed me. […] The problem was that he always had to be at home at the right time, I mean, I asked him to. That really was a bit annoying, like "Pablo, when will you get home?" or "Please, don't go out for dinner today" and so on.
>
> (Martina, 22 years old, four-time egg donor)

But indeed, the people who they relied on sometimes forgot or just disappeared, which caused fear and concern in donors. This was the experience of Lídia, who relied on her boyfriend to give her the shots.

> I never really jabbed myself because in reality I was scared. Well, it's something you need to be very mentally solid for. [...] Anyway, specifically that day [the day of the last injection], I don't know for what goddamn reason, my partner disappeared, and I had to do it on my own. And I had all the fears in the world of finding myself alone at the moment of truth. The following day, I had the procedure. I couldn't let air get in [into the syringe]. I wasn't sure if I was doing it right... maybe if I'd had someone with me, I would have been calmer. Someone who said, "Good, good, good, I think you're doing it right," and that's it, you know? It really was... It's cool to talk about it because I've never told anyone. It was like an evening that I spent doing that for an hour. An hour of nerves. I did the first shot wrong... I mean, I remember being so stressed... and maybe because... I also had inside a feeling of "I'm not sure what I'm doing," right? "But I have to do it because..." I found myself doing something that maybe... doing something without thinking, and then when you're in the middle of it you realize that you hadn't actually thought about it a lot. And well, I got it, yes. But the problem is that I was so nervous that I did it wrong... and in fact jabbing myself was very hard. I tried to jab myself, and I said, "Okay, I can't." I waited a bit, I lay down, but I had to do it one way or another. And I jabbed myself... Such drama, right?
>
> (Lídia, 27 years old, one-time egg donor)

The arrangements women have to make in order to medicate themselves correctly and on time are part of the labor of donorship. As Emily framed it, the required organization and the restrictions during the medication process—such as staying in town for examinations, not drinking alcohol, and not having sexual intercourse—made egg donation "like a job":

> It's like a job. Because when you go there, they give you a paper telling you about the indications on how you have to take the medication. And then you choose a time. You have a time slot, and you pick a time to take the medication. Of course, it was a bit limiting because it was Christmas, and sometimes I wanted to go out at night, but I couldn't go out because I had to stay at home... I mean, or I had to get home early, or I had to take the medication with me, or I don't know what, so that I could jab myself at the time that I had to. So, it was quite constricting in that way, right?

Moreover, for most participants this "job" overlapped with their other jobs, meaning that more decisions had to be made. Telling their bosses and colleagues about the donation or taking (or not) a couple of days off after the extraction were additional issues that they needed to resolve. Participants' talk about their choices reveals the vulnerabilities they faced in their work lives.

Juggling the Labor of Egg Donation with Work and School

Shortly after the incident at the train station, Mica had her extraction appointment on a Friday morning. Her mother drove her to the clinic and dropped her and her boyfriend two streets from the entrance. Walking that short distance, however, was harder than she expected: "I remember that we had to walk two blocks and that I could hardly move. I had to walk very slowly and I always… even if I was late today [for the interview], I'm always very punctual. Especially that day, I wanted to be punctual. My boyfriend had to hold me up because I was very, very bloated".

However, the extraction surgery was fast and soon she found herself in the hospital cafeteria eating "for free," with 1000 euros in her pocket and feeling "*super bien*" [really good]. "How did they give it [the money] to you?," I asked.

> In an envelope. I was thinking [laughs] it looks like I donated an organ. The girl told me, "Come here for a sec," and she took me to a small room and goes, "Here you go, next time it will be 1100 euros and the following 1200 euros." And I was like, "Calm down, I've just finished with this one." […] I remember that, on Friday, I thought "Well, if I continue like this [feeling this well], I'll do it more times," you know? And well, on Saturday, I was feeling a bit worse, and then on Sunday, it was the worst.

Mica started to feel bad the day after the extraction, and although clinics usually recommend resting, she went to work. "I should have rested from 24 to 48 hours as they recommend. But it was my second weekend of work at that hotel. And then, of course, I really wanted that job, and since I was going to be sitting, I said, 'I'll take the risk and go.' I worked from 8 am to 1 pm, until I couldn't bear it, and… I was in pain, bloated."

She was able to finish her shift on Saturday, but on Sunday she woke up and the pain was stronger. She reported to work at the hotel, hoping the pain would ease. But as soon as she arrived, she told her coworkers and boss that she wasn't well enough to stay. The matter was delicate because she had not told her boss the real reason for her absence on Friday. The only coworker that knew about the donation said things that made Mica feel bad: "It was like 'How can you do that with your body? You don't love yourself, whatever,' and well, I explained my reasons to her, and she said, 'How much do they pay you?' I said '1000 euros,' and she exhaled and said, 'Giving your body for 1000 euros. They have no shame, whatsoever.'"

The case of Mica is not unique. Although clinics recommend resting 24–48 hours after the extraction, most of the participants worked the next day, and some of them even worked the same day, on their afternoon shift. This was the case of Olivia, a three-time egg donor whom I met at the clinic during my observations and who worked at a supermarket. She came to the ovarian ultrasound check-up in the company of her 4-year-old son and stated to the doctor that she was '*molida*' [ground down] because of work. When I chatted with her again the day of the extraction, after a difficult recovery which involved loud groaning, she told me that at 4 pm, she was starting her shift at the supermarket. Participants went to work because they couldn't afford the lost wages and especially because they feared being fired if they missed a day.

Merche (26 years old, 19 when she donated) also went to work the same day of the ovarian puncture. As I explained in Chap. 3, she had been looking for a job "day and night" for five months and finally decided to schedule the donation cycle "to have a [financial] cushion." But just when she started the hormonal process, she finally found a job in a shoe store. "It must have been karma," she joked. The extraction was in the morning, and she had the afternoon shift:

> They recommended not to go to work that day, but I had just found that job recently… It was like starting the process and getting the job. I thought,

"Aren't I lucky?" [ironically]. And even more in a store. And precisely they scheduled the extraction on a Saturday morning. I thought, "How can I tell them, just after one week working there, that I'm not going to go on a Saturday afternoon because I have...?"

Luckily, she managed to be at the counter, where she could be sitting most of the afternoon, and she was able to complete her shift.

In many cases, egg donors chose not to inform their workplaces about the surgery to avoid possible problems. They hoped to be able to hide the pain and complete their shifts as normal. Their choices speak about precarious work relationships in Spain, which are characterized by a strong sense of insecurity. As I discussed in Chap. 3, we can also read from their stories the multiple struggles they faced to find work at their young age and how disposable they felt.

The findings of this section highlight that while the money received from the donation is legally conceptualized as "economic compensation" to cover the associated expenses and inconveniences, this is not how most donors used it. Rather than using the money to ease the process, most of my participants scheduled themselves as tightly as possible so that they could donate without disrupting their work life or incurring any extra expenses.

In the following sections, I focus on the management of pain and discomfort during egg donation, showing how this too requires labor from the donors.

Extractions: Fear and Loneliness at the Clinic

Paula, a nurse, opened the door when I got to the clinic on the day I was scheduled to observe extraction surgeries. We agreed that once the donors arrived I would have a few minutes to introduce myself and the research to them and ask for their consent to participate. The first one to arrive was Lola. We remembered each other from her ultrasound. After talking a bit about the study, I asked her if she was nervous, and she said she wasn't. It was the second time she was donating at this clinic, she told me. Very confidently, she agreed to allow me to observe the extraction and chat with me afterwards. Paula cut us off, saying it was time to go upstairs. She gave me some scrubs, and I quickly changed into them.

Shortly afterwards, the second donor arrived. I remembered her too from her ultrasound that I had observed a few days before. She was Olivia, the one who had brought her little boy to her appointment. Hers was actually the first extraction scheduled for that morning. She looked tired, like the day I saw at her check-up. She put her hands on her ovaries and told me that she was in pain. It was very hot in the room, and soon both of us were sweating. She agreed to allow me to observe her surgery. During Olivia's surgery, Lola would wait on a stretcher, all changed for the operation, in the room that served both as a waiting room and a recovery room for donors.

Once Olivia was ready, they quickly took her to the operating room and asked her to lie down on the stretcher. The anesthesiologist introduced himself and quickly explained what he would do while the nurse inserted an IV into her arm. The doctor came in and soon they started talking about the newly released movie *Aladdin*. The doctor highlighted the acting of Will Smith and reported that the little girl with whom she went to the cinema—her granddaughter I assumed—got sick afterwards. In the background, Freddy Mercury was playing on the radio. Olivia was looking at the ceiling, still, serious. She didn't smile or show any connection with the conversation. I thought about the fact that she had a 4-year-old son and that maybe she could actually participate in the discussion of the movie. No one looked at her and she didn't make any move to be included in the animated conversation. When they attached electrodes to her chest and her heartbeat started to be heard in the room, I saw on the monitor that her heart was beating at 114 beats per minute. "She's nervous," I thought. I didn't see the anesthesiologist put the sedative into her IV, but, as he would tell me later, in 16 seconds it made the loop from her arm to her brain and put her to sleep. The doctor put the ultrasound scanner and the needle through her vagina, and I saw the follicles on the screen. Looking closely, I saw how the needle went into them and sucked out the follicular liquid. The anesthesiologist talked about the incredible end-of-year party that was held at his daughter's school. "Not surprising that it was incredible, with all that you're paying!" said Paula (nurse), mocking him. The two embryologists in the room laughed and so did the doctor, while the anesthesiologist looked a bit disconcerted. As the anesthesiologist talked, he held Olivia's oxygen mask, almost completely covering her face with his hand. The biologists put the liquid that the doctor had sucked up into little dishes. In some of them, the liquid was more yellowish, in others more reddish, mixed with blood. Paula tagged the dishes with a black marker and put them in a container that kept them at a stable temperature. Later, in the lab, they would check them to see how many eggs they contained. And after a while, they would be able to see how many of them were mature, that is, suitable for IVF.

When the procedure was over, they switched on the lights and the anesthesiologist quickly took Olivia's mask off and started to call her name. In a second, she started moving. She opened her eyes and looked around the room. When her eyes met mine, I gave a hint of a smile. Shortly after, the anesthesiologist, the nurse, and the doctor, all at the same time, tried to wake her up by talking loudly close to her face. Olivia moved her head forward, as if she wanted to sit up, and the anesthesiologist pushed it down against the stretcher.

Finally, they took Olivia to the room where Lola waited. The stretcher passed in front of Lola as they placed it next to her, just separated by a blue plastic curtain. Olivia groaned. She said that she was very dizzy, that she was very sleepy. I thought she was going to cry. She looked at me in the eye and mumbled something that I couldn't understand. Lola seemed upset and inhibited. I asked Lola how she was feeling, "Fine, freaking out with this." Now, unlike before, she looked scared. I tried to reassure her, saying that this was probably a normal reaction to the anesthesia. But soon I remembered that it was not my role to make the clinic seem like a warmer place. At the same time, I wanted to keep her company, even if it was just by allowing her to share her feelings. Just like Olivia, Lola came to the clinic alone. After a few moments of silence in which I sensed she was trying to cope with the situation, she looked confident again and asked me to tell her whether she did something similar when she woke up from the anesthesia. She wanted to know what she would say under the effects of the sedative. I told her that if she said something, I would tell her.

The extraction is the moment of the process that aroused the most conflicted feelings, according to my participants. One thing reported, especially for first-timers, was the anxiety they experienced before the extraction. For some of them, the extraction represented a kind of awakening about the complexity of the process they were involved in. The common fear experienced before the operation became a realization of both how little control they had in the process and how vulnerable their bodies were. Nerves provoked bodily reactions such as stomach pain and uncontrollable shaking.

> I was… well I was like a *flan* [a custard dessert in Spain; the expression means "shaking like a leaf"], I mean super nervous. Of course, never in my life have I had any broken bones, never in my life was I hospitalized for anything, never in my life have I been in that situation. And the first time I was super nervous, it was 7 am. I was… with no breakfast, with nothing in my body, like a *flan*, but I mean shivering, right?
>
> (Sara, 25 years old, two-time egg donor)
>
> I was nervous… I was going crazy all alone, I was thinking "Oh my God, but what are you doing?" and they gave me a valium. I arrived there super high.
>
> (Arlet, 27 years old, one-time egg donor)
>
> I was very nervous, and between the nerves, the fact that you don't eat anything, your tummy makes noises… and of course, you go into a hospital, and I don't know if people feel the same way, but… the hospital smells like a hospital [laughs], all white, all doctors, all running… I mean, you find yourself a bit bewildered. But luckily a friend went with me, and you don't feel so alone.
>
> (Merche, 26 years old, one-time egg donor)

But while Merche highlights the importance of being with a friend that day, having company was not an option for all participants. Indeed, just as some donors went to work the same day of the extraction, some came and went to the surgery by themselves, against the recommendations of the clinic. I was surprised during my fieldwork to see that although the staff stressed the importance of donors coming with someone the day of the extraction, only one of the four donors I observed did so. The rest arrived at the clinic and left it unaccompanied. This was the case of Júlia (34 years old, two-time egg donor). Her main concerns with egg donation were not related to the risks or side effects but to the anesthesia.

> I started to link it [anesthesia] to the sexual abuse that I suffered when I was a child, and how they modified my behavior. It was like I always had to protect myself. Also, I had grown up in a family where my parents didn't take care of us the way they should have, and since I was a child, I had to take care of myself, right? It affected me for example in my adolescence, when my friends got drunk, I saw them as crazy, right? "How can people be here like this? Someone could do anything to them. Aren't they afraid of not being able to take care of themselves?" For me, losing control, for example, is something that makes me panic. So, one of the reasons [not to do it] was the idea that they had to sedate me and that I would lose this control. It's something that makes me panic at all the levels of my life. So, for this reason, I rejected the idea of egg donation, plus this sensation that they were going to extract my eggs, which was very related with the genitalia, you are asleep… This loss of control for me it was too much, right? And I let it go.

After some time, however, her long-time struggles with money worsened, and she decided to start the process. Despite having arranged with her sister to accompany her, the day of the extraction she found out that she would have to go alone.

> I talked to my sister, who long ago had said "Yes, I'll go with you," but the night before, she said she couldn't come. And I was like, "Oh my God, I have to go alone" and with my personal issue like, "Whenever when I have to do something that scares me, I'm alone." Like very obsessed with this: "Why is it that every time I have to do something that scares me there is nobody by my side?" So, I went in the morning, and I started crying as soon as I was out of my place, but crying… in the metro at 7 am, right? I went to the clinic and as I was closer, I cried more and more. And I'm quite sensitive and emotional but at the same time I'm like very cold, when I have to control my emotions I do, you know? And this thing of crying everywhere or that they received me there and I was crying, I mean that was like very strange, right? I arrived there crying and the girl that was there received me like "Oh, oh, come on, we'll go fast with you because I see that you're feeling terrible."
>
> **Anna:** Because they saw you like this?
>
> **Júlia:** Yes, of course. Plus, when I cry it's like I'm allergic to my own tears. My face looks terrible. […] And of course, they… here you see that

> what they're interested in are your eggs, because instead of checking what's wrong with that person, it's like, "Come on, we'll do it fast because you're having a hard time and we'll take care of it." Instead of seeing what's happening. And that's all. They put me in a room, "Put that on," and in the waiting area when I was alone, I cried. I didn't know why I was crying. Because well, I don't know, maybe also all the hormones that I had in me they were having this effect, and that's all, they put me in the operating room. And the doctor tells me I don't know what and she jabs me and so on. Which is quite a painful jab that one. And you fall asleep and that's it. That's it. You don't feel anything, you don't know what happens in that time, you don't know for how long you've been like that, and you start to wake up.

Once awake, however, the feelings came back again, and she cried some more. While she thought she was under the effect of the hormones, she was surprised to find herself still down and sad days after the extraction.

> After that, I was sad for a few weeks, if not more… but especially the first weeks very… very, very hard, I couldn't stop crying and crying and crying. And I took the bus, and I cried. I went wherever, and cried and cried and cried all day. Later it slowly eased, but that sensation… plus the sensation was of… I don't know, the questioning of what had happened during that time, and I was asleep, I don't know. Like a pile of things and… I don't know I was like very down, I cried a lot, I remember an experience like… (she exhales) I don't know.

For Júlia, the extraction was a very delicate subject, as she linked the anesthesia and procedure with the feelings of vulnerability she had experienced to the sexual abuse she suffered as a child. And although in Júlia's case, the association of the extraction with unconsented touching was clearly related to her previous experience, other participants also said they felt strange about being *remenada* ("messed with," a Catalan word that literally translates as "stirred up") while they were unconscious, which also evoked feelings of lost control and revulsion.

Laboring Discomfort and Pain

Similar to the way in which egg donation was framed as a simple process that makes the most of "all those eggs that would be thrown away by the body itself,"[4] bodily reactions and pain were minimized by clinic staff, who claimed that they were "like a period but stronger" (Núria, nurse). In this way, clinics naturalized discomfort and pain, making them easy to dismiss.

Waldby and Mitchell (2006) discuss how the discourse of "waste" is a way to disentangle tissue from the "personal relations that produced it, and from its significance as a potentially implantable entity, a possible future child" (p. 70). As they argue, this is a necessary process to turn tissue into bio-objects ready to enter the global circuits of biocapital. Similarly, Kalindi Vora (2016) highlights that the labelling of something (labor, tissue, or in this case, cells) as "unnecessary" *where* (and, I would add, *when*) they are currently located is a necessary operation before they can be mobilized and outsourced.

According to the clinics' discourse, pain, like eggs, goes to waste in every menstrual cycle. Egg donation, instead, gives a useful outcome to all these natural processes. This disentangling, therefore, erases not only the labor required by donors for its production, but also the sociotechnical and socioeconomic relations that make it possible (see Kroløkke, 2018; Neimanis, 2014; Shaw, 2017; Waldby & Mitchell, 2006). While most of the existing literature has focused on the construction of the eggs as "waste," I challenge the construction of discomfort and pain as a simple amplification of menstrual bloating and cramps.

As a challenge to clinics' discourses, I consider the management of discomfort and pain to be a type of labor that egg donors perform to become bioavailable. Unlike any form of "natural" pain due to menstruation, egg providers have to endure new forms of physical and emotional pain when undergoing an egg extraction cycle. Mica's story of giving herself a shot at the train station provides a compelling example of how the subjugation of the body for egg donation can lead to unexpected strong bodily reactions. Other participants had similar, though milder,

[4] IVI clinic's website.

experiences, in which unexpected bodily sensations disrupted their carefully planned disassociation from their bodies:

> I didn't like it at all, not a bit, not a bit… and now that I'm thinking about it, maybe I would have liked some more preparation for when you wake up, the sensation that you have. Because for me… it was 15 minutes, right? Apparently, it's just a moment, but they keep telling you "Now you will fall asleep," and I think that I wasn't aware that I really wouldn't be aware of anything, you literally fall asleep, and I woke up feeling that I had been fully stirred up inside [*remenada* in the Catalan original]. And I didn't like it at all. I remember that they put like a diaper on me, you know? Like a very big pad. They stirred you up and you're losing a bit of blood… and suddenly you say "Wow, it hurts a bit," just a tiny bit I mean, right? And I didn't like that. Because… right? It's like when you're asleep and think, "Wow, someone touched me, and I didn't realize." But they can't do it any other way. And I remember that that sensation was very weird. "This morning someone has been touching me, on my [sexual] parts, and I wasn't… I was aware but I didn't notice."
>
> (Arlet, 27 years old, one-time egg donor)

Sara described a feeling inside her vagina as if someone "took a condom, filled it with ice and put it inside of me."

> They didn't hurt me or anything I mean… but what I did feel during the extraction was a very cold sensation here in my vagina, I mean, inside. Not on my clitoris or… it was inside. As if… I don't know, I was asleep, because I was asleep, but I was like dreaming that I had a super strong, cold sensation. As if they took a condom, put ice inside it, and put it inside you… I mean, a sensation of cold that almost burned. I mean, that was very strange, the first time it didn't happen. I don't know if it was that the anesthesia didn't have a strong effect on me, I don't know. And when I woke up, I told the doctor, "Wow, I felt a very strong cold." He told me that they put in some helium… helium, is that possible? To open up the area and not leave a scratch… to be more careful.
>
> (Sara, 25 years old, two-time egg donor)

One of the most striking cases was that of Andrea, who told me that before the extraction she looked 7-months pregnant, because of bloating:

> Well, after this, it's true that I noticed a lot of hormonal problems. Afterwards, moreover… I'll look for the picture because I must have it. My belly became like a 7-month pregnant belly. Two days before, I couldn't move. I had a very hard time during the process. But horrendous, right? A belly that… impressive. I had a very bad time. But well, of course, the needles, you have to jab yourself every day. Obviously, it was my mom who did it, because for me it was impossible. […] You have sensations… I guess that like when you're pregnant, of course I've never been pregnant, but you have cravings, the feeling that you are very full, very heavy… you also feel very strong hormonal waves, like ups and downs.
>
> (Andrea, 29 years old, one-time egg donor)

Andrea's comparison of her bodily experiences to those of pregnancy was not just merely a metaphor, as can be seen in pictures she shared with me after the interview, in which her belly is swollen as if she were pregnant (Images 5.5, 5.6, and 5.7).

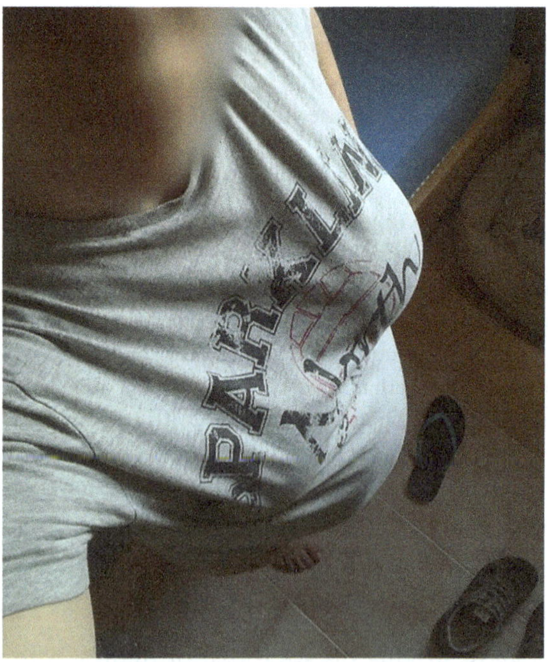

Image 5.5 Andrea's selfie of her belly during the medication process (I)

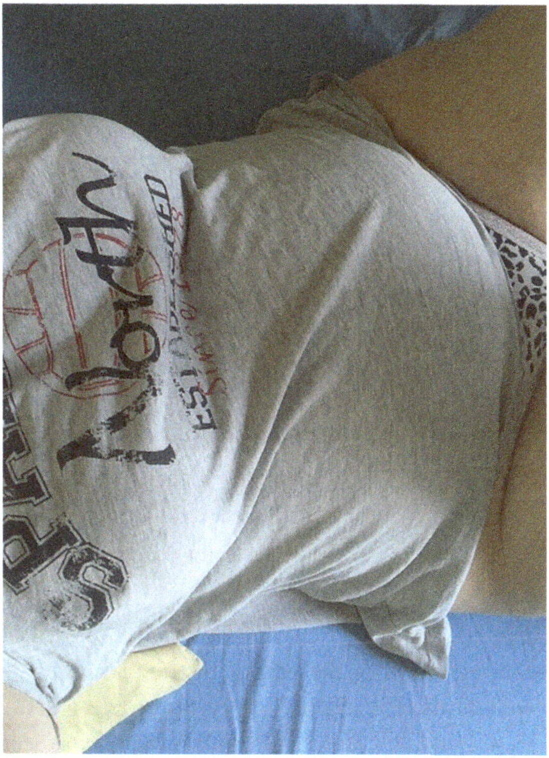

Image 5.6 Andrea's selfie of her belly during the medication process (II)

These examples also provide clarity on a common issue for many participants: the difficulty of making sense of their bodily reactions. As I argue in the following chapters, this visceral way in which participants expressed resistance—through their bodies—is closely related to the discursive and material practices of the fertility clinics, which worked to separate the donors as biographical subjects from the donors as body objects.

Conclusions

In this chapter, I have discussed the different operations required for intending egg donors to make themselves bioavailable, claiming that they are all part of the labor of donorship. First I have examined the process of

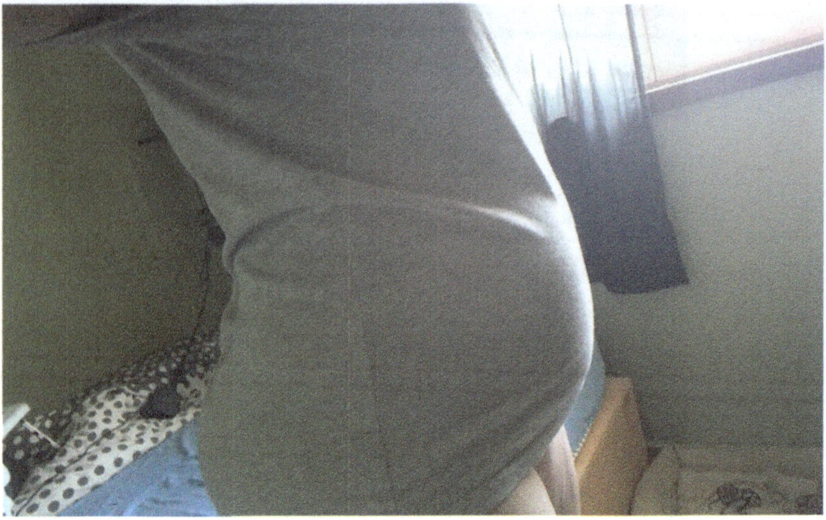

Image 5.7 Andrea's selfie of her belly during the medication process (III)

choosing a clinic to undergo the extraction cycle, showing that donors engaged in the imaginaries of the brands of the private fertility clinics in Barcelona as well as their aesthetic aspects in order to assure safety and professionalism. Choosing clinics that are renowned was seen as a way to feel safe amid the perception that some clinics may not perform ethically and may be more dangerous. However, I have argued that while choosing a clinic was the first step for intending egg donors to assert agency and control, many of them quickly realized that they might not be able to choose. In this part of the process, donors started to understand clinics' requirements, and therefore, their own biocapital in the fertility industry. Participants came to understand how physiological aspects such as weight and height as well as phenotypical features affected their potential value in the eyes of the clinic.

Next, I have analyzed the arrangements that intending egg donors had to make to complete the process. Here, I have focused on how they managed their self-medication and made arrangements to meet the demands of work. I have described the difficulties they encountered in managing the injections and attending their appointments while also meeting other commitments. I have highlighted that most of the egg donors tried to meet the demands of egg donation without affecting their work and/or studies. Further, I have noted that while the economic compensation is legally conceptualized as a way to offset the possible loss of income, egg donors didn't view or manage it this way. Rather, they tried to prevent spending any of it to ease the process and instead treated it as additional income. Finally, I have focused on the pain and discomfort egg donors endured throughout the process and how, contrary to the discourse of fertility clinics which describe it as similar to those of a menstrual cycle, women undertook exceptional amounts of emotional and physical labor before, during, and after the extraction. I contend that undertaking these arrangements involves a process of subjectification and taming, which facilitates egg donors' alignment with clinics' extraction purposes.

Describing the labor of donors is a crucial task under the growing interest in reproductive bioeconomies. This examination has the potential to reveal not only its different forms, but also the conditions in which it is undertaken. This labor is currently made invisible through the discourse of altruism and the anonymity of donation in general discussions of the growing reproductive bioeconomies that ignore such labor's central role. I argue that it is imperative to materialize through thick ethnographic detail the labor that goes into becoming a subject for the expanding bioeconomies in order to make visible the role of the (often gendered) work of reproductive laborers as a source of value creation.

In the next chapter, I shift the gaze from egg donors to fertility clinics and clinicians to examine the selection practices and taming techniques used to control donors' behavior.

References

Brown, W. (2015). *Undoing the demos: Neoliberalism's stealth revolution.* Zone Books.

Cohen, L. (2005). Operability, bioavailability, and exception. In A. Ong & S. J. Collier (Eds.), *Global assemblages: Technology, politics, and ethics as anthropological problems* (pp. 79–90). Blackwell Publishing.

Cooper, M., & Waldby, C. (2014). *Clinical labor: Tissue donors and research subjects in the global bioeconomy.* Duke University Press.

Kroløkke, C. (2018). *Global fluids: The cultural politics of reproductive waste and value*. Berghahn Books.
Nahman, M. (2013). *Extractions: An ethnography of reproductive tourism*. Palgrave Macmillan.
Namberger, V. (2019). *The reproductive body at work: The South African bioeconomy of egg donation*. Routledge.
Neimanis, A. (2014). Speculative reproduction: Biotechnologies and ecologies in thick time. *PhiloSOPHIA, 4*(1), 108–128. https://doi.org/10.1353/phi.2014.a549347
Pavone, V., & Goven, J. (Eds.). (2017). *Bioeconomies: Life, technology, and capital in the 21st century*. Palgrave Macmillan.
Perler, L., & Schurr, C. (2020). Intimate lives in the global bioeconomy: Reproductive biographies of Mexican egg donors. *Body & Society, 27*, 3–27. https://doi.org/10.1177/1357034X2093632
Rudrappa, S. (2015). *Discounted life: The price of global surrogacy*. New York University Press.
SEF. (2022). *Registro Nacional de Actividad 2022*. Sociedad Española de Fertilidad.
Shaw, R. M. (2017). Bioethics beyond altruism. In R. M. Shaw (Ed.), *Bioethics beyond altruism: Donating and transforming human biological materials* (pp. 3–31). Palgrave Macmillan.
Siegl, V. (2023). *Intimate strangers*. Cornell University Press.
Teman, E. (2010). *Birthing a mother: The surrogate body and the pregnant self*. University of California Press.
Tober, D., & Kroløkke, C. (2021). Emotion, embodiment, and reproductive colonialism in the global human egg trade. *Gender, Work & Organization, 28*(5), 1766–1786. https://doi.org/10.1111/gwao.12637
Vora, K. (2016). *Life support: Biocapital and the new history of outsourced labor*. University of Minnesota Press.
Waldby, C., & Mitchell, R. (2006). *Tissue economies: Blood, organs, and cell lines in late capitalism*. Duke University Press.
Warren, N., Canaway, R., Unantenne, N., & Manderson, L. (2013). Taking control: Complementary and alternative medicine in diabetes and cardiovascular disease management. *Health (London), 17*(4), 323–339. https://doi.org/10.1177/1363459312460699
Weis, C. (2021). *Surrogacy in Russia: An ethnography of reproductive labour, stratification and migration*. Emerald.

6

Containing Behavior

Introduction

One day, as I was leaving a fertility clinic after an interview with Laura, its lab director, she took the opportunity to introduce me to the reception staff, comprised of Núria and Paula (nurses), and Roser (administrative assistant), whom I would interview the next day. "They are the ones who suffer the donors," Laura said, and the three of them smiled ironically and nodded. "*Suffering them* is a good way to put it," Laura reaffirmed.

Indeed, professionals across the different clinics described egg donors as difficult to work with and staff members stated—often in a joking tone—that they needed patience, the ability to take things in stride, and skills such as deep breathing and meditation to handle this apparently very stressful job.

> Then what is really important is what they're like, right? Because if they're not responsible… there is a very irresponsible profile of donors and that's chaos, because if they don't medicate themselves correctly… We don't classify them as good or bad, or beautiful and less beautiful in general, but in terms of responsibility. A chick who does a good donation, but then two bad ones because she hasn't medicated herself correctly, then another good

A. Molas, *Taming Egg Donors*, Health, Technology and Society,
https://doi.org/10.1007/978-981-96-5103-0_6

> one… This is terrible in terms of numbers. It's crazy in terms of numbers and of emotional cost for the recipients, because of all that it involves, a donation that doesn't work out is chaos, right? Both in terms of management and in terms of the patient.
>
> (Montse, lab director)

As Montse put it, an applicant's level of responsibility was a crucial consideration in the donor selection process. The clinic's fear that it might not get a return on the investment in tests, drugs, and time made crucial the work of the staff, especially those who had the initial encounters with donors, such as reception staff, nurses, and psychologists.

In her influential exploration of the ways in which "medicine enacts the objects of its concern and treatment" (p. xii), Annemarie Mol (2002) describes the different practices that transform a single object into more than one (multiple). Drawing on her fieldwork in a Dutch hospital, the author focuses on the different coexisting ontologies around atherosclerosis enacted by the different actors who—from different standpoints and with different aims—interact with disease.

In my observations and interviews with the staff of fertility clinics, I learned that the bodies of donors were also multiple. Two main ontologies of donors emerged as stable across different clinics and professionals. The site of exploration where they were the most evident was in the selection and screening process that intending donors underwent to enter the egg donation program. As the aim of egg donation for reproduction purposes is to achieve a pregnancy, it would be expected that selection practices in fertility were directed toward selecting fertile bodies able to reproduce marketable phenotypes while meeting health standards. While this is true, a range of other important factors in the selection process emerged in my fieldwork, revealing an ontological duality, in which egg donors-as-subjects and egg donors-as-objects are screened and tamed in different ways. In this chapter, I focus on the enactment of egg donors-as-subjects, while I discuss donors-as-objects in Chap. 7.

As many clinicians highlighted, the demands of the egg donation process (attending all ultrasound check-ups; self-injecting the hormonal medication daily for 8 to 10 days; abstaining from sex, alcohol, and drugs during the process; enduring pain; and persisting until the day of the extraction surgery) made it very important to assess aspects beyond the reproductive capabilities of egg donors' bodies. Selecting women who were deemed responsible was as important as ensuring that their eggs would have a commercial output.

In this chapter, I provide insight into the ways women were read, selected, and managed in the clinical setting, and I explore what these discourses and practices tell us about broader Spanish biopolitics. I first describe the contexts in which the donor as biographical subject emerges. Providing fieldwork examples, I discuss the problems that clinicians dealt with when treating egg donors and how imaginaries of class and race intersected in their rationalities. In this endeavor, I demonstrate that egg donors were perceived as having an "excess" of reproduction that clinics redirected to parents and families deemed as more desirable for reproduction. Second, I focus on the criteria clinical psychologists used to select women perceived as responsible and ultimately likely to follow the rules and bring the process to completion. Here, I delve deeper into the intersections with class and race and explore the importance of women's motivations in this assessment. Third, I show that in addition to strict selection protocols, clinics had mechanisms in place to assure that donors brought the donation to completion once they started, which I discuss as taming techniques directed to ensure trouble-free cycles. In this category, I distinguish between taming techniques directed to *control* donors' behavior through coercive methods, from the taming techniques based on the *care* of egg donors, which ultimately were directed to create intimacy between the donors and the clinic.

In an earlier article, Laura Perler and I examined these multiple donor bodies emerging in fertility clinics to highlight the rationalities underpinning selection practices and taming techniques (Molas & Perler, 2020). In this chapter, I flesh out and supplement some of the concepts used for that earlier work, where we proposed using the terms *biographical subject* and *body object*.

We used the term *biographical subject* to refer to the donor enacted as a subject with agency that has to be tamed throughout the donation process in order for the cycle to be completed. We linked this term to the concept of "reproductive biographies" (Perler & Schurr, 2020), which is used to highlight the entanglements between the global bioeconomy, state biopolitics, and women's intimate lives. With the term *body object*, we referred to the donor enacted as a site of extractability. With this concept, we pointed out the coded conversations that take place between donors' bodies and medical experts which, as I will unpack in Chap. 7, are not translated to the donor as a biographical subject.

Tracing the Desirable Donor: Class, Race, and Reproductive Waste

Like Montse, interviewees Paula, Núria, and Roser also highlighted the great burden of working with egg donors. Staff members painted a picture of the average donor, a woman who was irresponsible and untrustworthy, traits that staff members linked to their social class and country of origin. Staff described the hardship of their job as dealing with the messiness of egg donors' lives—or more accurately, of the lives of people of a particular social stratum.

Paula: The first issue is whether they come.

Roser: Whether they attend.

Núria: Punctuality above all, because out of ten, eight aren't punctual. [...] Or they say "Yes, I'm leaving home now, I'm coming and so on" and then she doesn't come, and you think, "Well then don't tell me [that you're coming] and that's it." It's a bit that you don't really believe what they say.

[...]

Paula: Let's see, the ones who need and the ones who offer their eggs, they are already two different profiles of people, I mean, a donor... I'll just come out and say it, okay? They are... in most cases, okay? They don't have much education of... of... I mean, they are a segment that's very... I don't know, who had a bit of a difficult childhood, a complicated adolescence. They are not very stable financially. [...] [After describing the steps of the first medical check-up] if everything is inside the normal parameters, she continues with the process. Which would be to give her all the documents that Roser was mentioning, informed consent [forms]...

Roser: Which they don't even read.

Paula: Which they don't even read, exactly.

Núria: Well, a bit broadly...

Roser: The QCarrier[1] they don't read it for sure.

Paula: We give her an informed consent [form] to do a genetic analysis, and many of them don't read it, or they don't fill it out because they don't read it, or they sign it directly and sometimes they don't even put their names on it.

Núria: They have lots of pages on genetic diseases and so on and come on, they don't read it for sure.

[…]

Paula: And afterwards, we also give them the psychological test, which many don't pass either. Because they don't read it properly… It's what we were saying. They do it fast or they have a family history or personal issues that you think, "No wonder they don't pass it," or because… their IQ, poor things, is… a bit low for understanding certain questions […]. Of course, also the level… What I was telling you, the intellectual level is important. When you see that they're a bit lost in thought, you think, "She won't do it right because she's not understanding anything." I'm telling her this, but she's not understanding anything. I mean, you see them with a poker face like saying, "What the hell are you talking about?," you know? I mean, not even if I explained it with pictures, right? She's not understanding anything. And of course, you think, "Doing a treatment with this person doesn't make sense, because… it's really unlikely that she'll do it right." The truth is that most of the time when they do it wrong it's because of this. They fail because they couldn't make the solution, "I forgot to take the medication that was due," "I was…" whatever, "I was daydreaming, and I didn't stop taking the [birth control] pill…." One thousand stories because of what we were saying, they don't seem engaged. They think like, "I'm going to get a perm at the hairdresser and that's it," and you say, "No, no, no, it doesn't work like that." So, many say, "I can come that day" [mocking high pitch voice] "because I'm free…" and you say, "This doesn't work like that. This is not 'I'm going to have my manicure done.'"

Disorganized, with complicated backgrounds, financially unstable, and with low levels of education/IQ were some of the ways that Paula (nurse) described most of the women that came to the clinic to donate. During my observations with them at the reception desk, I was able to see a clear example of the way they read egg donors and how social class played a

[1] Carrier screening genetic test.

crucial role in it. Two young women entered through the door of the clinic one morning. They were from outside of Barcelona, but I didn't catch the name of their town. One of them was coming for an ultrasound check-up, since she was already taking the medication, and the other to receive the results of the genetic carrier screening, which would determine whether she could start the stimulation cycle. The one that was already in the medication process told the staff that just before starting with the injections, her period was late, leading her to think she might be pregnant. She told the story of her pregnancy scare in great detail (in the end she wasn't pregnant). Núria and Roser, remaining serious, listened to her story in silence. Once both donors had been sent to the doctor's consultation room, Roser looked at me with a sarcastic expression and said, "These [women] get pregnant as often as I go to the supermarket."

The fact that she was highlighting how easily and irresponsibly "these" (low-class) women could become pregnant was significant in the context of a fertility clinic. The boundaries of the desirable way to reproduce were suddenly clear. While the sexuality and reproduction of lower-class women was conceived as surplus, involuntary, and excessive, the reproduction of patients in fertility clinics—who were middle-to-upper class—was rather framed as a blessing, a way to make dreams come true, and to "make families" in a controlled and responsible way. The ideas about the desirable subject to be reproduced in this setting are clearly entangled with conceptions of class, and as seen in the following paragraphs, also of race.

Conceptions of race and the migrant status of many intending donors were also brought up by Paula, who was convinced that there were differences between them and local egg donors. Migrant women from Latin American and Eastern European countries (the vast majority of all foreign egg donors according to all the clinics I spoke to) were perceived to be irresponsible not only in their personal lives, but also because they became egg donors "without thinking too much," which they mocked even though this was useful for the clinic. Ideas about the easy and careless ways these women reproduced appeared again in their narratives as they drew on colonial imaginaries of migrant women as hypersexualized, extra fertile, informal, and lazy.

Paula: There are a lot of Latinas. Most of them have a mindset of, "There's nothing wrong with being an egg donor." They are Latinas… Latin Americans are like, "I donate this like I donate blood…." They don't have… especially because of course, they are mothers much younger than Mediterraneans. Then it's like, "Look, I'm a mother of two, of three, of four, whatever, and now I donate eggs and help other people and, moreover, I am financially compensated." It's like… they have this much more internalized than Mediterraneans, right? Caucasians are like, "What do you mean donate eggs? What does it involve?" Right? They are like more reticent. There's a culture that's not… don't you think?

Núria: There are also a lot of Russian donors.

Roser: I think it's because what we're saying, it's for the money. Many come here and they don't have papers to work so they need income. They need money somehow. And some of them are more capable of thinking and so on… but others, they go directly to this, right? […] Of course, until they get their papers, they can't find a job and so on. So, the ones that are from here, generally they work or they're in school, then it's different, right? Because their intellectual level is different, the level of education they might have is different, and all the status they have. […] But these ones [the migrants], it's another world, another level, I mean, they don't even think, they just go with the flow. According to the circumstances… not of life, but of that same day, on what they have, on what happens to them, "Well, I'll do that, I'll help someone. They'll give me money, and, in the meantime, I'll look for a job, or I'll go to the beach."

Since the first wave of migrants came to Spain in the early 90s in a government strategy to cover jobs that the country could not fill with its population, migrant women—mainly from Spain's former colonies—have been the visible face of domestic work in Spain (Marre & Gaggiotti, 2004; see also Marre et al., 2018). Migrants, many of whom are undocumented, have been crucial to supporting the country's fertility rate, which would otherwise have plummeted well below replacement levels. Despite this key contribution and similar to Milena Marchesi's (2012) findings in her study of reproductive biopolitics in Italy, migrant women are scrutinized and mocked for their excessive reproduction, which "is construed as a problematic and even dangerous contribution to the nation" (p. 173). In this context, the clinic becomes the site for transferring reproduction from less desirable subjects to more desirable ones.

These racist and classist dynamics tie into the discussion around reproductive waste from Chap. 5. The clinical discourse frames egg donors as wasting their reproductive capabilities, their cycles, and their eggs, which become revalued and redistributed in fertility clinics to reproduce other more desirable values, couples, family ideals, and a specific social class more broadly. In the following figures (Illustrations 6.1 and 6.2), I compare the different imaginaries around egg donors and recipients according to the clinicians I interviewed.

Donors were perceived as subjects that embodied the risk of unreliably exerting their freedom and agency by, for example, deciding to drop out or by committing mistakes in the medication process. The problem for clinics, which they openly recognized, was that their business model

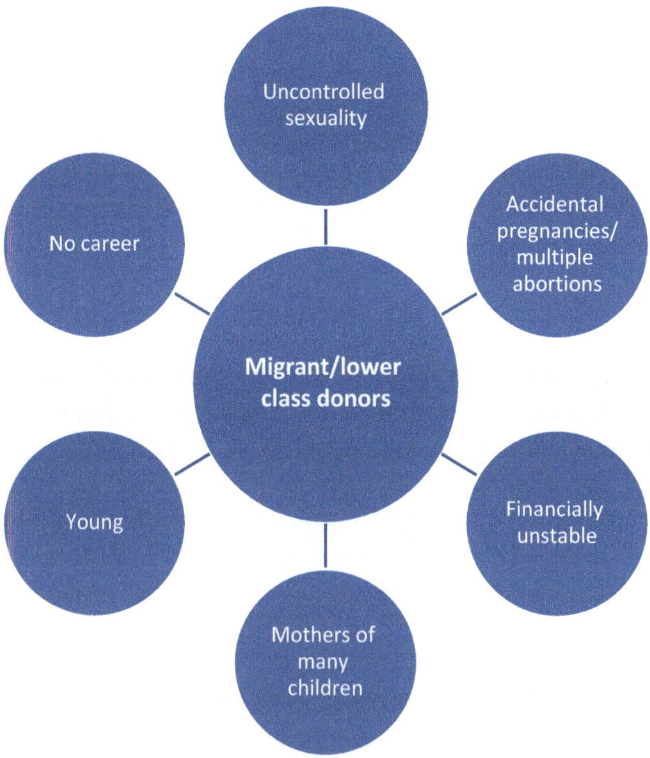

Illustration 6.1 Imaginaries of donors according to clinicians

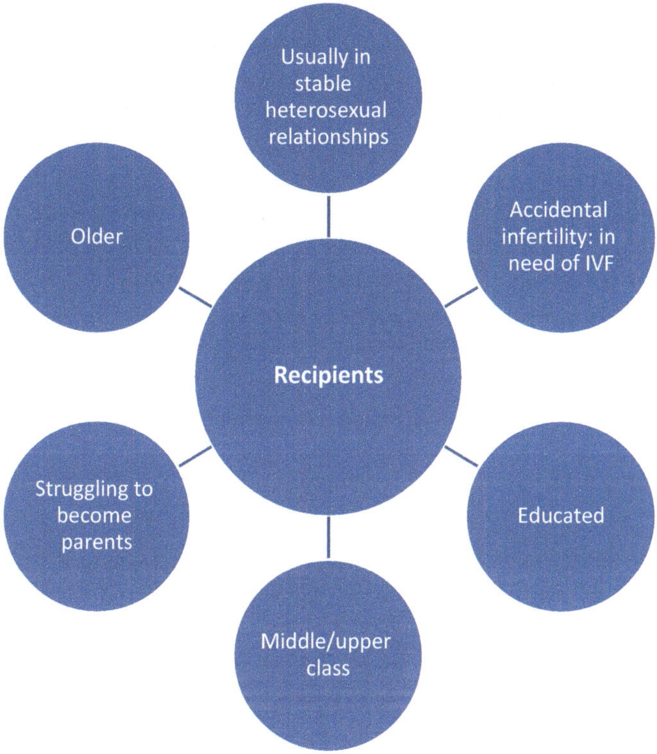

Illustration 6.2 Imaginaries of recipients according to clinicians

relied on donors. Therefore, treating, negotiating, and taking care of them was viewed as a great burden but one from which they could not escape. At least *not yet.*

> Those who have this bit of irresponsibility, well, not irresponsibility but lack of responsibility, it's very difficult for you to modulate, they're like this and that's it. But what happens? Sometimes they have uncommon physical characteristics, blond, blue eyes, green eyes, so we're interested even if they're messy and… you're interested in doing another cycle with her even if she gave you headaches because she wasn't on time for the visits and so on.
> (Aina, gynecologist)

> Moreover, the donors know that the clinics need them, right? So, this also relaxes them a lot. It's like, "Well, if you don't do it, I'll go to another place." And unfortunately, many times they're right. I mean, you have to put up with it because you need her.
> (Laura, lab director)
> They go to the one next door. It's exactly like that. Now with the whole thing about SIRHA[2]... SIRHA is the platform that theoretically since... The law says that the donors should be monitored and now they're not monitored. And the SIRHA won't allow you to be at two clinics as a donor. Because now it's like trading. When we started the clinic in Barcelona, it was chaotic, like, "What will you give me?" "But I want to start next month because otherwise I'll go to the clinic next door" and so on.
> (Montse, lab director)

Aina emphasized that the degree of patience that staff members had with a given egg donor was related to the desirability of her phenotype, highlighting again the relevance of biocapital among egg donors. While many clinicians had the impression that egg donors took advantage of knowing that they were so highly needed by clinics, clinicians also acknowledged that there was nothing they could do to escape from having to deal with them. Interestingly, some of them believed this dependence would eventually come to an end (see Epilogue).

Selecting Responsible Biographical Subjects

Given this "chaotic" pool of women, selecting appropriate donors became one of the most crucial tasks for fertility clinics. One interesting figure of the egg donation screening procedure in Spanish private fertility clinics was the clinical psychologist, who in most clinics had at least one formal interview with the egg donors before they were accepted into the program. As two psychologists explained to me, their function was twofold: on the one hand, they had to determine if the donors were at risk of

[2] SIRHA is the abbreviation of Sistema de Información de Reproducción Humana Asistida (Information System of Human Assisted Reproduction), which is the name of the donors' registry that was in its trial version at the time of my fieldwork.

transmitting hereditary mental illnesses. And on the other hand, they had to make sure that donors would act obediently throughout the process, medicate themselves correctly, and attend check-ups.

> It is especially interesting to see her life habits, right? What they do, if there's emotional stability, work stability… if there's social stability, right? Sometimes we see a lot of people who come from abroad, women from other countries come here to donate. Maybe they are women that have been here only a few months and because of their economic need they look for something to get money. And they come here, then we have to see how their adaptation to society is. Where they live, if they have friends, if they don't… I mean, there are a lot of things that you can see in the person, right? But above all, we want them to be honest, okay? Be honest. That really the life that they have is the healthiest possible, okay? That she has an interest in life. Sometimes they are mothers, so check the care that she is giving to her little ones. Sometimes they are girls who are separated, who are coming out of a process of separation, and then we check how they're handling this separation from their husbands. Sometimes we have people who have been physically abused, right? So, we check how she was able to get out of that, to look for resources to get out of that by herself from this situation. Above all, what we look for is to see people… well, with a certain emotional balance and with an adaptation to society. Their health and nutrition habits as well, does she exercise… are they active in their free time… We also look for people who are active, right? And who are healthy. (Gemma, psychologist)

Adaptation to society, an active social life, emotional balance, exercise, a healthy diet, and being a "good" mother were identified as indicators of the overall commitment an intending egg donor would have during the egg donation process. When highlighting a special assessment for migrant women who donated for economic reasons or women who had been physically abused by their partners, psychologists, and clinicians more broadly, implicitly provided a picture of the dangers associated with being from a specific social class.

> It's true that you see the vibe of a person, right? So, if for example she's a person who doesn't have any kind of rules at home… who has been in and

> out of jobs because she's been fired, because she didn't arrive on time. Or she didn't follow the rules, who maybe has sex every week with a different person. Of course, these are signs of people that don't have the responsibility that we have to demand for such a treatment, because she doesn't commit to anything, right? This commitment… You see the commitment, you see it, you see the commitment she has in her life, in her goals, and you see that she doesn't commit to anything. The friendships, she might have some but well, if it goes alright, great, and if it doesn't, that's great too. Maybe her family isn't so important to her, right? These are little things that make you see that she won't be able to stick to this commitment. Or they tell me, "I wake up and I go to the park, then I go to a cafe and then maybe I clean a house, but if I don't want to, I tell them that I'll come tomorrow." That means that this personal commitment is not there.
>
> (Gemma, psychologist)

Unemployment or precarious employment (such as sporadic domestic work) and a general "lack of motivation" in life—traits highlighted as undesirable—imply a preference for donors with economic stability versus women who are at risk of social marginalization. As already highlighted by Sara Lafuente-Funes (2017), clinicians in Spanish fertility clinics "show a clear preference toward having donors that behave in particular ways: altruistic, calm, responsible, and discreet" (p. 264). Indeed, to find bodies able to produce eggs through trouble-free cycles, the selection process must identify subjects that are malleable, obedient, and responsible. Motivation is one important variable. As implied in the preceding quotations, women perceived as being only interested in the money are perceived as being more likely to quit before completing the process. The connection between an intending donor's behavior and her motivation (that is, whether she is driven more by the compensation or the desire to help) is in turn perceived to be linked to her social class.

> **Raquel:** In that first interview… the most important thing when they come here, the first thing is honesty. Honesty. Afterwards, obviously, well, if they look you in the eyes, what she's telling you… why they came to donate, the why. […] But I think that the first thing, in the first minute, the most important thing is to see the response that she has, what she's looking for, how she looks at you… There are many women who come and

> the first thing that they tell you is that they have family members who had fertility problems or "My sister just had a baby and I thought that I could pass on this happiness," right? "And I'm young, and I can pass it on." Then well… you're already seeing that there's this family thing, of generosity… that they want to help. […]
>
> **Anna:** Why is it important that they want to help when they come to donate?
>
> **Raquel:** Yes, it's important because… it's true that you can find people who are less altruistic and that care more about the money. If they come very, very eager to help, the process that they have is also more beautiful. Calmer. Their insides are also more relaxed, the endometrium is like more relaxed [laughs].
>
> **Anna:** Really?
>
> **Raquel:** I can't assure you because I haven't done a study, but they are more respectful when it comes to coming to the clinic, their punctuality… They bail on you less in the appointments. Because they have this idea that they want to help and their goal is to help, they feel good doing this. So, for us, it's important maybe as a facilitator, right? Also, for all the processes that they will undergo, and because of the fact that they want to help… It's not that it confirms it, but it gives you an idea that they will be like very honest.
>
> (Raquel, psychologist)

Irene, the egg donation coordinator of the same clinic where Raquel worked, made a similar statement:

> It's like in real life, there are hunches, right? For example, last week there was one that had something that I didn't like… You can't really say what it is, right? They don't look you in the eye when they speak, you feel that they're thinking of something else, like they're not paying attention. At the end of the day, you're explaining to them some risks, although it's true that they're not huge, but you need to be aware, right? […] You see that they aren't paying attention, or you see that they want to get to the point and say, "No, I'm not so interested in what the surgery will be like," and they ask you directly about the compensation. […] This doesn't necessarily mean that she won't do the treatment with us, but you take it into account when you… you say, "Well, this girl can be used to do a deferred treatment or for egg freezing," so that she doesn't have a recipient waiting, right? At the same time, you try to control it.

As Irene stated, the results of this team assessment of the egg donors in terms of their background and social status did not always result in the rejection of the donor, but sometimes it determined the type of treatment they received. If a donor with desirable phenotypical traits for the clinic displayed unwanted behaviors or troubling hints about her life, they might choose to do a cycle to freeze all her extracted eggs and put them in the egg bank, instead of using them for a fresh cycle in which a recipient has to be synchronized. This method assured that there would be no consequences for the recipient should the donor drop out of the process halfway or make a mistake in taking the medication.

Taming Biographies Through Control and Care

In addition to the efforts of clinicians to select women likely to have smooth and trouble-free cycles, clinics had two sets of taming techniques in place to address women's behavior and performance throughout the process. In the following subsections, I delineate the two sets of techniques: those that were directed to *controlling* egg donors and those that were directed to *caring* for them.

Control: Tests, Money, and Medication

"Hooking donors" with Fast Medical Screenings

First, as stated in Chap. 4, donors who expressed interest in the donation after the first informative appointment were often given the informed consent form and the medical check-ups that same day. As different clinicians confirmed, this was part of a strategy to motivate donors to come back to the clinic a second time.

> When they come to be informed […] there are many clinics that, before they go home, they already give them a blood test. Because this is a way to make them want to come back to know the results or to find out about her blood type. It's a way to hook her. So, in the second visit, you give her an

ultrasound scan, or even in the first visit… The thing is to do everything fast, so they don't disengage and so that they want to know the results.
(Magalí, embryologist)

Threatening to Withhold Payment

A second widespread way to manage donors' behavior was by threatening to reduce the compensation or withhold it altogether. In this regard, different clinics were observed to perform differently following their own protocol. While some paid donors only if they got to the extraction, others paid a small amount if a cycle had to be canceled partway through for medical reasons (such as her body not reacting as expected to the stimulation medication).

They are sacred [the ultrasound check-ups] but sometimes also… there are some who don't come, right? But… it depends a bit on the person. If the person is a bit flaky, you have to keep after them. A lot. And at the end, the sentence that motivates them the most is, "If you don't come, you won't receive the compensation," and then yes, many of them come back quickly, you know? Then you say, "Okay, so the donation was altruistic, right?" [laughs], but well, that's the way it is.
(Laura, lab director)
Keep in mind that the compensation isn't given until the eggs are extracted. Until you've been through the operating room, they don't give you anything. Because, if at the first visit, they give you 200 euros, everybody would come, would take the 200 euros and "goodbye." No. So, until you get to an operating room, or if they can't extract the eggs for medical reasons, you aren't paid. Even if you've done everything. I mean… they make you hold on a lot.
(Magalí, biologist)
We give a partial compensation because she didn't make it to the end. I mean, what is compensated for are the inconveniences, so of course, the main inconvenience is the extraction, apart from the daily injections. Depending on the moment when they are canceled, we have a protocol that… obviously if it was a negligence on her part, I mean, that she hasn't taken the medication, that she has lost it… then the compensation is minimal or can even be non-existent if she stopped coming to the [ultrasound]

check-ups. […] And well, it depends on the moment when the process is canceled. If they do it all correctly and they are committed people and they didn't do it on [purpose]… then it's almost complete. It depends a bit.
(Neus, gynecologist and clinic manager)

Reimbursement

Third, donors were not just threatened with the possibility of not being paid, but also with the possibility of having to reimburse the clinic for the cost of their treatment should they want to leave the process halfway through. As stated in the template of the egg donation contract offered by the SEF, "[I]f the extraction does not proceed because of withdrawal, negligence or bad faith of the donor, the center will also be able to demand that she repay the cost of the explorations and treatments that might have occurred."[3] Although the professionals I spoke to stated that they had never enacted this clause, it is possible that fear of being forced to reimburse the clinic led some women to disappear altogether if they faced difficulties in the middle of the donation process.

WhatsApp

Fourth, clinics used WhatsApp to monitor egg donors in ingenious ways. At the most basic level, they used WhatsApp's read receipt feature (the double blue check mark) to know whether a donor had read their messages.[4] They also used donors' WhatsApp profile pictures to conduct an initial phenotypical screening, through which intending donors with "very pronounced Latin American features" might be ruled out before the interview (Núria, nurse, in an informal conversation in the clinic. More on how racial selection worked in Chap. 7). One clinic also reported that communication via instant messaging was useful to monitor whether and when the last medication (the one to trigger ovulation) had been injected. This was done by asking donors to send a screenshot of their

[3] Original text in Spanish. Fragment translated to English by the author.

[4] The read receipt function can be turned off in WhatsApp, but by default it is on.

phone at the moment when they gave themselves the shot, as proof of the time.

During my observations at the clinic, Núria, Paula, and Roser let me look at the smartphone they had to communicate only with egg donors. Most contacts were saved like "Leticia, *amiga de* Rosana" [friend of]. When I asked Núria about it, she told me that knowing from whom that donor came from gave them a lot of information about what to expect. If the recommender had been a "good" donor, they felt more secure about the new one. It was also useful, they told me, because if a donor disappeared in the middle of the process, they could always get in contact with the person who recommended the clinic to ask what was going on.

Stimulation Drugs

Finally, donors are sometimes managed through the medications. Corifollitropin alfa (distributed under the commercial name of *Elomva*) is a long-acting stimulation drug that can substitute up to six days of daily medications with only one shot that is injected at the clinic. As the price for this medication is higher than that of the other drug (which requires daily injections), it was usually reserved for donors who were perceived to be more irresponsible and those with highly desired phenotypes for which the clinic did not want to risk a cycle cancellation due to misadministration.

> There's a medication that's the most expensive and that not all donors can have, but it's one shot and that's it. After that, they need to have more injections, but at least we would do this first collection fine. And this one, they give it to them here [at the clinic]. Then many times you say [to the doctor] "Prescribe her this one because I think it will be the only way to do it correctly."
>
> (Paula, nurse)

Virginia, a biologist and the coordinator of the egg donation program of a fertility clinic in Barcelona, also pointed out the increased benefits that the implementation of this medication provided:

> So, with this hormone, you're already avoiding mistakes for five days. [...] So, the hormone diminishes... In fact, since we started using it, the percentage of bad responses has lowered a lot. Before we had to rule out seven, eight percent [of potential donors]. Now we have to rule out two percent of donors, okay?

Care: "Here we treat them like princesses": Affective Capital in the Private Clinic

When I started knocking on the doors of different private clinics in Barcelona, I quickly got the impression that my fieldwork plan was going to be difficult to accomplish. This was not unexpected. However, the narrative behind the clinics' denials was surprising. While I thought that the main problem clinics would have would be related to the fear of donors potentially speaking critically of the clinic and perhaps of me seeing and hearing about the precarious background of most of them, this generated less discomfort than I expected. As I described above, most medical clinicians spoke openly about the obvious economic motivations of most of egg donors as well as the fact that, for example, a large number of them were migrant women from Latin America and Eastern Europe. However, every time a potential contact with donors was raised, they referred to them as very delicate subjects.

Before starting my fieldwork, I had contacted different fertility clinics to discuss my research project and I got the informal agreement of one of them. When I visited the clinic in person for the first time to give clinicians a copy of the different documents explaining the project and the survey I wanted to conduct with egg donors, Neus, gynecologist and head of the clinic, said that although she was okay with everything I wanted to do, she would have to discuss it with the rest of the team. I was a bit afraid that they would be surprised to see that many questions in the survey were directed to understanding the socioeconomic position of women, but I was confident that the clinic would agree to participate. However, a few days later, Neus called me. She was clearly in a different mood, more distant. She told me that the psychologist didn't like the research project and that they would not let me give the survey to donors myself, nor talk to them outside the clinic, nor, obviously, be in the examination room with them. "Here we treat them like princesses," she said, "like our daughters." And then she highlighted numerous times that out of the number of girls who called for a first appointment, only around 30% actually came at the end. She also explained that donors often expressed annoyance and an attitude of "stop bugging me" [*no me marees*], leading clinic staff to go out of their way to avoid bothering them.

Only a few days after that, I had a very similar conversation with the coordinator of the egg donation program of another fertility clinic. "Well, the donors are a group that…" and she exhaled looking at the ceiling. Then she told me that there were a lot of couples, local and foreign, who came to the clinic for egg donation treatments, and that this was why they "took great care of the donors." She told me that studies that were very demanding of egg donors usually didn't pass the clinic's approval committee. She was telling me all of this with silences, breaks, and taking her time to look for the appropriate words. She told me that the vision in the clinic was that of "Let's not bug the donors" [*a les donants no les maregis*]. I was very surprised that she used the same words as Neus to explain why my study could not take place in their clinic. It became clear to me that their main fear was not so much what I could see or hear about the donors but rather the possibility of them dropping out if the process became too tiresome.

From their explanations, I came to understand that clinics considered their relationships with donors carefully, and they didn't want to do anything that might jeopardize their willingness to donate. In my observations and interviews with both practitioners and egg donors, I realized that other equally important taming techniques were taking place at the clinic which were instead based on the investment of a significant amount of affective capital in egg donors.

"At a time when there are more than 30 private fertility clinics in Barcelona, how can you make sure they will go to your clinic?" I asked clinics' staff in our interviews. The answers I got from my participants, who were mostly practitioners at medium-sized clinics, were usually in the same line as that of Dolors, a nurse who had been in charge of her clinic's egg donation program for years:

> Big clinics have the advantage that they have a great economic capacity, and they can do a lot of advertising through radio, TV, on [the sides of] buses, and you see it everywhere. Instead, small clinics, we don't have this possibility. For us, a donor must be treated really well [*super bé*]. You need to make it as easy as possible for her and make sure she feels taken care of. So that when they come, they know who they'll find, and they'll see that you know their name and that she's treated in a very personalized way. That she's not a sheep, okay? She's not a number. And we tell them that if they liked it and they want to tell other people or... and the truth is that there are girls that come and say, "Yes because so-and-so told me and...." I suppose that it happens at other clinics as well, but for us this is important.
>
> **Anna:** The fact that they come because a friend recommended you?
>
> **Dolors:** Because a friend did the treatment here, yes. In fact, there are girls that tell you that they did donations in other places and then you ask them why they've changed, and it's because they weren't happy at the other place. They tell you, right? And you say, well so even more reason to spoil them, take care of them and put them at ease...

"Spoiling donors" and making it "as easy as possible" materialized in a set of different practices that clinics had been developing throughout the last 10 years, as more and more clinics were established and the competition for donors became tougher. These taming techniques were directed to managing the perceived fragility of egg donors' will as well as the fragility of their life more broadly. These practices reveal the gendered care labor required by front-line staff, the explicit economic dimension of the notion of taming, and the particular kind of relation that results from it.

Women Carers as Capital

Dolors' clinic, for example, had hired one specific young woman, Irene, to deal with issues related to egg donors, and reported that "the investment," as they themselves referred to it, was a success. Gerard, who had been the main gynecologist of another private fertility clinic for years, told me how hiring women to do these caring tasks was not uncommon:

> There are places where there is specific staff member who deals with the donors, I don't know, women doctors, more like girls, young, cooler… trying to make the ambience less medical. Even if, in reality, you're doing a medical procedure. So that the patient perceives it like, "Well cool, this isn't too serious," like, "I feel comfortable, and this isn't too, too, too serious." Like a middle point between being professional but not too serious. And yes, it's usually outsourced to a specific department.
>
> (Gerard, gynecologist)

This investment in care for donors was well received by some participants in this research, who, regardless of the concerns they might have about the industry, usually described the care they received in positive terms. Alexia, for example, expressed how the fact that she was going to be treated by women was an important aspect of her decision-making.

> And well… it was always quite good. I guess that the fact that they were women influenced me, because if it was men, it would have been more… because the person that answered my email was a man, his name was Jordi and I felt like, "Watch out." But then they told me, "We've assigned you a woman doctor." Actually, almost all of the nurses were female which, like it or not, has an influence. And I don't know, they were all very kind like… this kind of people that are very cute like "ooh…" [using a sweet soft voice], and that also… the personality of people makes you want to come back or not.
>
> (Alexia 21 years old, one-time egg donor)

Clinics generated chains of capital from women based on emotional (Hochschild, 2003), embodied (Fannin, 2020), and intimate labor (Boris & Salazar Parreñas, 2010). Female nurses or coordinators were in charge of caring for donors, who were asked in turn to care for themselves during the process in order to care for the recipient woman waiting for their eggs. The interplay between clinics' efforts to retain donors and donors' own delicate negotiation of ambivalence highlights both the fragility of the egg donation market and the specific gendered labor and relationality involved in making it work.

Continuous Communication

A second investment clinics made to care for donors was to maintain intensive and continuous communication with them throughout the process. While this included phone calls from time to time, in most cases donors were given a phone number to contact the nurses through WhatsApp. In my observations at the clinic, I saw that they had a specific android cell phone that they used exclusively to talk to egg donors. As mentioned before, the use of WhatsApp allowed clinics to do more than exchange practical information with donors; it also became a way to provide donors with a sense of proximity and security:

> I've always had the same nurses and the same doctors. At the end, I was going there, and they said, "Hello Martina, how are you doing?" The ones from the secretary's office, they knew me as well, and bear in mind that, in the morning, there were different people than in the afternoon. I mean, yes, the faces were the same. It's not like in a hospital where there are many doctors, right? There it was them and I knew them and… they also do something that makes you feel very close, which is that they give you… well, they have a WhatsApp number through which they ask you, "How are you feeling after the extraction? Have you already had your period? How do you feel?" This also makes it very warm, the fact that you see that there's like a real follow-up also made me feel that I wasn't putting my body at risk.
>
> (Martina, 22 years old, four-time egg donor)

Very few egg donors I interviewed had ever experienced private health care before they became egg donors. As I discussed in Chap. 5, most of them were middle to lower class and not used to such levels of personalized care in medical settings. The young age of many of the participants at the time of the donation meant that, for most of them, the extraction cycle was by far the longest medical intervention they had ever had, with some of them even stating that it included their first appointment ever with a gynecologist. Additionally, in some cases, donors hid their donation from their families, making the clinics' investment in affective capital especially meaningful.

For example, Merche decided to donate eggs when she was 19 years old. She had just declined her admission to a public university because she could not afford the fees. After five months of unsuccessfully looking for a job, she decided to do an egg donation cycle (more details on her case in Chap. 3). Her mom, who had raised Merche and her brother alone and was still supporting them with long shifts in the kitchen of a hospital, was not aware of her daughters' decision. Therefore, Merche managed the injections alone in the bathroom and hid the medication in the fridge pretending it was her allergy medication. The exhaustive care and attention from the clinic helped her to cope with it all:

> It's not the same now at 26 than then at 19, I'll tell you that. When I found myself with the needles, in the bathroom… plus my family didn't know anything. Because maybe their mindset wasn't like mine. This is the first thing. And also, maybe they would think, "She's doing this for money or why is she doing this? She has other options…" or things like that. Then I hadn't said anything, and it was a bit… I found myself at 19 years old in the bathroom with the syringes and I was like, "Okay… what am I doing?" [laughs]. But once you do it the first time, you get the hang of it, plus during all this time your *médicas* [female doctors] are always in contact with you, they call you, you go, they also do some gynecological appointments in the meantime. They ask you during the process how you're doing with the medication, how you are, whether you feel alright… So, there is always contact. It's not like they give you the drugs and forget about it. There's always contact between the clinic and you.

Medical Care Benefits

A third care investment involved performing small favors to current or recent donors such as providing medical services for free during their time as donors. As Neus, gynecologist and clinic manager, stated, clinicians sometimes saw themselves as helping women from lower social strata to successfully navigate health services.

> **Neus:** Even after [the extraction], if a donor tells you, "Oh, I need to get an IUD put in. Could you do it for me?" I say, "Get it, get it and I'll put it

in." I mean, we do… we try to follow up because some of them come from some social strata where they struggle to get an appointment at the outpatient clinic and so on, and well, "So if you go to buy an IUD, we'll put it in." I mean, lots of times we continue with the medical service afterwards.

Irene: There's support beyond the donation itself.

Neus: If we find an abnormal pap smear, we insist, "Have you been to the outpatient clinic? Have you been there to check it out?" We try to perform the role of a primary gynecologist while they're doing the process or….

Irene: Or between donations, yes.

Neus: Now, if it's been three years since they donated and they come to me with an IUD, then maybe not, right? But while they're in the process of… we consider them to be patients and we have to take care of them.

(Neus, gynecologist and clinic manager, and Irene, egg donation program coordinator)

Another example is free ultrasounds for recent egg donors. This was the case for Maria, who found herself accidentally pregnant a month after the donation (case study in Chap. 5). Maria called the clinic, and the staff offered to perform an ultrasound for free.

Specific Staff for "Russians"

A fourth way in which this care materialized was in the resources in place for migrant donors, especially Russians who might not be proficient in Spanish or English. The clinic where I did observations had the egg donors' informed consent documents in four languages: Catalan, Spanish, English, and Russian. That same clinic also had a specific person who was exclusively in charge of recruiting and accompanying Russian donors through their time in the clinic. The day I observed egg extraction surgeries I got the chance to meet her because one of the donors who was undergoing the extraction that day was Katia, who, Paula (nurse) had warned me, was Russian and not very proficient in Spanish.

I met Katia in the waiting room where she was waiting alone. She was blond, slim, and pale. I started introducing myself and explaining my research when she stopped me, gesturing to indicate that she was having a hard time understanding. I continued much slower and asked her from time to time if she was following. While talking to her I found out that she was not actually from Russia, but from Ukraine, although she let me know that she spoke Russian. Given her basic level of Spanish, I assumed that Katia had arrived in Spain recently. But to my surprise, she said that she had been in the country for 4 years. She told me that, as she spent most of her time at home caring for her 4-year-old son, she didn't have many chances to practice Spanish. When I asked her how she was feeling, she struggled to find the words in Spanish but finally told me that she felt "nauseated." It was the third time that she had undergone an extraction cycle, but she didn't plan to do any additional cycles: All the hormones that she had to inject herself with scared her, she told me.

Soon, a woman wearing nursing scrubs came into the waiting room. It was Masha, the clinic coordinator in charge of the management of Russian patients and donors. She said something to Katia in Russian, and in response Katia stood up and went to the room next door to get changed for the operation. When Katia's extraction was over, Masha entered the recovery room, sat next to her, and started talking in Russian. At some point, Masha explained to me that she tried to stay with the "Russian" donors during their entire stay at the clinic, although she abstained from being in the operating room, as once she fainted in the middle of an extraction. She couldn't stand blood. As Masha kept talking to Katia, I wondered if this role existed at other clinics. And I pondered how someone like Katia, who barely spoke Spanish and barely left the house, ended up at the clinic.

I followed them when Masha took Katia downstairs to give her the final documents and the compensation. Roser (administrative staff) gave her 1000 euros in 50-euro bills inside an envelope. I could see it because they made her count it before putting the envelope in her bag. Katia left the clinic alone. Masha then went to her office, and I took the opportunity to ask Roser if she knew how Katia ended up in the clinic. "Some Russian mafia," she said earnestly. Then she told me that she didn't really know and that the Russian coordinators probably had contacts, mentioning the "Russian mafia" several more times. She told me, as if trusting me with a secret, that she found it strange that sometimes donors "come to the clinic just one or two days after arriving from Russia."

A bit later that same day, a young woman came into the clinic and soon Masha came down again to receive her. Just like with Katia, she stayed close to her and started talking to her in Russian. The new woman had come to pick up her medication. In the time that she was there with Masha and Roser, who gave her a plastic bag with the drugs, the only thing that I heard her say in Spanish was *adiós* [goodbye]. A bit later I had the chance to talk to Masha for a few minutes. Again, I asked her how someone like Katia arrived at the clinic to donate eggs. She told me that there were many donors who came "by recommendation" and that there were also networks of Russians who lived in Spain to which the coordinators of the clinic advertised themselves, mainly through social media. She told me that her presence at the clinic was a guarantee for Russian donors that everything would be alright and that the doctors wouldn't hurt them. She explained that, "of course," because many donors came to the clinic without speaking the language, they were afraid that the doctors would take their organs. She also told me that because "economic issues" were donors' main reason for donating, they told each other about the clinic and came "by recommendation."

As many clinicians told me, having Eastern European donors was a great asset for a fertility clinic. It allowed them to provide phenotypes featuring light hair and eyes, which were in great demand and more difficult to cover with the Spanish donor pool. Having staff whose job was exclusively to establish links between Russian-speaking women and the clinic is an example of how fertility clinics used care work to compete for egg donors.

Part of this investment in care for donors materialized in the hiring of women staff whose role was precisely to create a more calm, caring, and comforting experience. This practice was directed to reassuring the donors that the risks were under control or that, as Masha put it, no one would take their organs. This work was especially important when considering the backdrop of egg donors' experiences. As discussed, egg donors were usually very young and may have been doing the donation in secret and/or in a country foreign to them.

Segregation in Space

A fifth way in which clinicians cared for egg donors was through the separation of the spaces between egg donors and recipients. This practice has been increasing in recent years as many big and middle-sized clinics have opened specific centers for egg donors. Some other smaller clinics have built different waiting rooms and, in some cases, even different doors to enter and exit the clinic. While some clinicians stated that separating egg donors from recipients was a way to make them both feel more comfortable and to maintain their privacy, space segregation also allowed clinics to accomplish other goals. The segregation of spaces allowed clinics to maintain their image of professionalism and luxury, which was challenged by the uneasy presence of donors, who often embodied markers of lower class and age.

> **Gerard:** Well… there are donors that I think, "If the recipient saw her, she wouldn't want her." Not physically, because she might be very cute or look very similar [to the recipient]. But because… she's very sloppy or full of piercings or… for example, smoking habits, drugs, these are not exclusion criteria. The fact that a donor smokes is not exclusive. There are donors who smoke like a chimney and so on, who are party animals, who arrive late to the appointments because they come from partying and so on… mmh… and sometimes there are recipients that you think, "Wow, if the recipient saw this donor she really wouldn't want her." She thinks the donor is a perfect *bomboncito* [sweetie] who studies at Cambridge, you know? And sporty, and healthy, and so on. And it's not always like that. And sometimes there are donors that I think, "If the recipient saw her, she wouldn't want her." But not physically… I mean, she accomplishes objective medical criteria, she accomplishes them. But in terms of behavior or… well.
>
> **Anna:** Do you think this is why contact is avoided a bit?
>
> **Gerard:** Yes, contact is avoided because of that, so that they don't judge them and so on. So, the donors don't feel uncomfortable either, so they're in a friendlier environment and so on, and so that the recipients… Well, because as the most frequent indication for the treatment is age, it's very frequent that at the waiting room people look at each other like: "You're a donor," "You're a recipient," so that no one feels uncomfortable, because it's

> very clear who's 44 years old and who's 19. In a waiting room, it's like you see it clearly… you would see the 19-year-old chewing gum, smelling of cigarettes, sprawled out in a corner and so on… I'm exaggerating, but sometimes it's like that, right? And the 44 year-old, I don't know, super posh and so on, reading the newspaper… and then it would be really noticeable. […] Because a lot of donors that… the clinic doesn't want *that* to be what you see in the waiting room.
> (Gerard, gynecologist)

Through the practice of splitting donors and recipients into different spaces, the clinic has the chance to maintain a blurry imaginary of the donor as a distant young woman who is responsible, well-intentioned, intelligent, and healthy. Removing the donors from the central clinic is yet another strategy to remove eggs from their present and their past and to control the images that each party forms of the other.

Appealing to Donors' Feelings and Moral Obligations

Finally, both my observations and the interviews with egg donors showed how clinics tried to manage egg donors by creating close relations that were imbued with moral obligations. In my observations at the clinic during the ultrasounds, for example, it was interesting to observe gynecologist Valentina as she talked to egg donors. While she did not give them explicit information on the progress of the cycle, she bolstered them with sentences such as "You're a genius," "You're doing so well," terms of endearment such as *cariño* (honey), *corazón* (sweetheart), and a joking name for the medical gown that left the backside uncovered: *faldita hawaiana* (little Hawaiian skirt).

This had consequences in the way egg donors navigated their journey at the clinic. Arlet, for example, told me that when she was having her first informative appointment, the doctor almost broke into tears after receiving a call which, as she explained to Arlet, informed her that after a long process a 47-year-old patient had received a positive pregnancy test. "I don't know if she was acting," admitted Arlet, "but it worked." Similarly, Alexia told me that for her last donation, the clinic had contacted her via WhatsApp to tell her that a donor had dropped out and that they needed a replacement as soon as possible. They highlighted the pain of the recipient who was already waiting for the eggs. She agreed. After the second donation was done, she received messages from the coordinator of the program three more times, which she found too persistent:

> I don't know, the insistence was a bit like… stop it, right? And I don't know, I guess this is the thing that was the most… I mean, I didn't… I didn't feel obligated to do it at any time or anything, everything was because I wanted, but well… Maybe I felt that the insistence was a bit too much when I had already told them no, I told them that I was going to go abroad. Which was a lie, obviously, but well… I said it because… I don't know.
> (Alexia, 21 years old, two-time egg donor)

Faced with the insistence of the clinic, Alexia finally decided to lie so they would stop texting her. The fact that she said that she was going abroad instead of reiterating that she did not want to do it anymore illustrated the emotional strings attaching donors to clinics. Another example was that of Raquel (psychologist) who, as evidence that her clinic took good care of donors, told me that some donors would run up to give her a hug when they saw her in the clinic hallway.

> They hug you and say, "This is so good, I'm happy!" And they're undergoing injections! Right? And they say, "And you know what they told me, Raquel? They told me that from the last one [the last donation], the baby is born, he's born!"
> (Raquel, psychologist)

This image of the ideal donor-clinic relationship (a donor running through the corridors to embrace the psychologist to celebrate a birth) gives an idea of the level of emotional involvement that was desired from donors. Clinics hoped for and cultivated donors who committed to the cause and the clinic on a personal level, whom the clinic could trust, and who could be counted on if there were any problems.

While clinic staff members said that the rapport they built with donors was intended to make donors feel comfortable and safe, it was also used to win donors' cooperation by appealing to their empathy. Indeed, clinicians sometimes told donors about their problems or successes at the clinic, embedding them in highly emotional narratives, which in turn made it difficult for donors to say "no."

Moments of Realization

While, as I have shown, many donors thought that the relationship they had with the clinic was very good, some of them also described moments at which they felt like they were waking up from a pleasant dream tailored by fertility clinics. These feelings usually appeared when they faced health complications, when the process failed, or in particular situations they observed at the clinic. The examples below—while by no means typical of all donors or all clinics—give voice to some of these experiences.

Diana, who started to feel a lot of pain the night after the extraction surgery, called the clinic to let them know that she had a fever and that she was feeling unwell. She was told to get checked at a public hospital:

> Then, the next day, I had a fever. I called the clinic because of course, I had a bit of a fever, and they told me to go the Hospital Clinic [public hospital in Barcelona], you know? Because they didn't want to take responsibility. I told them that I wouldn't go, that they did it [the surgery] and that I was going to the [fertility] clinic. I told them that I wanted the gynecologist at the [fertility] clinic to check me. So, then they gave me an antibiotic and so on. The gynecologist that examined me was the emergency one and of course, he didn't have anything to do with the assisted reproduction unit,[5] and he was like shocked, you know? Like a bit, "Well, this could be… you can have an ovarian torsion, you know?" I mean… that I could have problems afterwards, right? So, I think they [the clinicians] did it [the extraction] pretty rough, I don't know.
>
> (Diana, 24 years old, one-time egg donor)

Diana's story brings to our attention another interesting private/public dynamic also discussed in Chap. 4. When there are complications derived from the treatment that require a fuller examination or hospitalization,

[5] The fertility clinic that Diana attended was connected to a private hospital, which was where she was sent after refusing to go to the emergency department of the public hospital.

patients were told to visit a public hospital. So, while the procedure was performed by private fertility clinics and the profits were distributed to them, the costs of health complications were usually assumed by the public health system. As several clinicians told me, this was the case not only with egg donors, but also with fertility patients, who usually required further specific medical assistance derived from the risks that an in vitro fertilization pregnancy in a person over 40 years old inevitably involved.

In these situations, the participants often felt that the care they had received through the process was opportunistic. As Emily explained, after having experienced a lot of pain following the extraction, she felt that the final call from the clinic was insufficient:

> Then I had a very bad time with the anesthesia because it made me feel terrible. I had nausea and I spent all day long vomiting. Then they call you, and that was the last call I received from them [smiles ironically]. It's like very pragmatic, right? While you're there, they'll call you as many times as they need to, until you perform your function. But once you've performed it, [it's as if the clinic says] "Now I don't know anything about you." And that was the last call. They asked how I was feeling, if the anesthesia was wearing off and… that's it. They haven't called me again.
> (Emily, 23 years old, two-time egg donor)

Another participant, Berta (21 years old, two-time egg donor), also had a moment of realization at the end of her second donation. The gift they offered her along with the economic compensation made her grasp the different ways her body was at work in the fertility clinic. She shared her experience during a focus group with two other participants: Alba and Lídia.

> **Berta:** I went there [to the clinic] for the last time in January. I mean, it was very surreal. They gave me the money and afterward they told me that because I had done it [the cycle] around Christmas… they gave me a voucher for free fake eyelashes.
> **Lídia:** Oh, what are you saying?
> **Alba:** Really?
> **Lídia:** Together with the money, right?

Berta: I mean, they asked me if I wanted it, and I said no. And then I was like, "Well, give them to me...," but I mean...

Alba: I thought you were going to say, I don't know, a free medical check-up...

Berta: A voucher... for fake eyelashes.

Alba: [There's] nothing more feminine, do you realize?

Berta: They're there, in an envelope at home. I don't know if it has an expiration date. [...] I haven't even opened it, I mean, I don't want fake eyelashes! I mean, exactly, if it was a voucher for a medical check-up or... I don't know man, it doesn't matter [they all laugh]. A spa, a voucher to go to a spa to relax. Look, this would have made more sense to me, right? After all that a donation involves. Fake eyelashes...

Alba: And well, I don't know if it was a coincidence or not, but it's very sexist that they give you precisely this.

Berta: Exactly, yes, yes, yes. It's super sexist, it's like, "wow" I mean...

Alba: Like, "You're a woman, you have to like fake eyelashes, for sure."

Receiving a voucher for fake eyelashes was perceived by Berta, Alba, and Lídia as a kind of humiliation. To them the gift suggested that the clinic viewed donors as superficial, materialistic, and stereotypically feminine, a point that recalls dismissive statements by nurse Paula (cited above) about donors who supposedly thought of donating eggs as similar to getting a perm or a manicure. The gift also perverted the narrative of the donation, leading them to feel objectified. Furthermore, establishing this link between the reproductive industry and the beauty industry (which in the case of some clinics is literal)[6] disconnected assisted reproduction from the realm of the medical and instead connected it to the realm of body enhancement. If providing eggs enacted their femaleness in a narrative that highlighted solidarity among women, one thousand euros cash in an envelope and a voucher for fake eyelashes enacted their femaleness as well but in a different way: the body as a site for aesthetic enhancement to highlight femininity through materialism. For Berta, the gift was a kind of betrayal: The clinics' previous acts of care were replaced with a reminder of her position in the fertility industry.

[6] Eva Clinics, a rapidly expanding Spanish fertility group, belongs to the Grupo CC which also has plastic surgery clinics (Dorsia) that usually share space and part of the staff.

Conclusions

In this chapter, I have argued that an ontological duality of egg donors is enacted in fertility clinics. I have distinguished between the *biographical subject* and the *body object*, following Annemarie Mol's conceptualization of multiplicity (2002). This chapter has focused on the delimitation, the selection, and the taming practices of the biographical subject, for which I have referred to the donor enacted as a subject with agency that had to be contained throughout the donation process in order to make the body available for extraction.

Dealing with egg donors was experienced as challenging by clinicians, who often viewed them as unreliable due to the complicated personal situations and backgrounds of many of them. The imaginaries of class and race that were intertwined in staffs' perceptions of donors highlight, by contrast, the contours of the desirable reproducible subject. While many egg donors were labeled as disorganized, financially unstable, uneducated, and irresponsible in terms of their sexuality and reproductive patterns, the clinic became the place to transfer this "excess" of reproduction toward more desirable parents, who were usually defined, in opposition to egg donors, as middle-age, middle-to-upper class, educated, and responsible in terms of family planning. In this context, selecting women perceived as reliable and likely to complete the cycle was a crucial task.

Despite the efforts of clinicians to assess the level of responsibility and commitment of donors, biographical subjects were always perceived as being at risk of abandoning the process or causing cancellations. Therefore, clinics had taming techniques in place to ensure that donors would continue to the end of the process. The techniques can be separated in two groups, those directed at *controlling* egg donors and those directed at *caring* for them. Control practices included fast preliminary screenings to "hook" donors, economic measures to tie compensation to compliance, routine monitoring through WhatsApp, and medication protocols aimed at reducing mistakes. Care techniques, in contrast, included appointing specific staff, usually young women, to treat and coordinate egg donors and maintain continuous communication to create intimate and emotional ties with donors, offering favors such as gynecological health care in between cycles and segregating donors and recipients in different spaces to help donors feel more at ease and also to obscure their markers of class and age from recipients. On a final note, I have noted the experiences of some egg donors that felt that these care relationships broke down at some point, revealing the donor's unequal positionality within the fertility industry.

By describing the techniques in place to tame egg donors' behavior, I have highlighted how taming operates as a dual and relational process of control and care that is key for the extraction of eggs for global reproductive circuits. The perceived excesses of egg donors' vitality—infused with class, race, gender, and colonial imaginaries—are constructed as obstacles for cycle completion and position egg donors as both as sites of value extraction and subjects of fragility. These taming techniques address not only the perceived unpredictability of donors' will but also the fragility of their lives more broadly. These practices reveal the affective capital invested by fertility clinics that translated into gendered care and intimate and emotional labor by front-line staff, highlighting the explicit economic dimension of taming practices as well as the ambivalent relation of control, care, and mutuality that emerges within the fertility industry.

In the following chapter, I focus on the donor enacted as *body object*, selected for its capacity to reproduce health standards and desirable phenotypes and tamed so that clinicians could have intimate conversations with it, conversations that were not translated for the *biographical subject.*

References

Boris, E., & Salazar Parreñas, R. (2010). Introduction. In E. Boris & R. Salazar Parreñas (Eds.), *Intimate labors: Cultures, technologies, and the politics of care* (pp. 1–12). Stanford University Press.

Fannin, M. (2020). Embodied labour in the bioeconomy. In A. Datta, P. Hopkins, L. Johnston, E. Olson, & J. Maria Silva (Eds.), *Routledge handbook of gender and feminist geographies* (pp. 326–335). Routledge.

Hochschild, A. R. (2003). *The managed heart: Commercialization of human feeling*. University of California Press.
Lafuente-Funes, S. (2017). Egg donation in the making: Gender, selection and (in)visibilities in the Spanish bioeconomy of reproduction. In V. Pavone & J. Goven (Eds.), *Bioeconomies* (pp. 253–277). Palgrave Macmillan. https://doi.org/10.1007/978-3-319-55651-2_11
Marchesi, M. (2012). Reproducing Italians: Contested biopolitics in the age of "replacement anxiety". *Anthropology & Medicine, 19*(2), 171–188. https://doi.org/10.1080/13648470.2012.675043
Marre, D., & Gaggiotti, H. (2004). La inmigración hispanoamericana: Relaciones, estereotipos y realidades. In S. Carrasco i Pons (Ed.), *Inmigración, Contexto Familiar y Educación. Procesos y Experiencias de la Población Marroquí, Ecuatoriana, China y Senegambiana* (pp. 95–128). Universitat Autónoma de Barcelona, Institut de Ciències de l'Educació.
Marre, D., San Román, B., & Guerra, D. (2018). On reproductive work in Spain: Transnational adoption, egg donation, surrogacy. *Medical Anthropology, 37*(2), 158–173. https://doi.org/10.1080/01459740.2017.1361947
Mol, A. (2002). *The body multiple: Ontology in medical practice*. Duke University Press.
Molas, A., & Perler, L. (2020). Selecting women, taming bodies? Body ontologies in egg donation practices in Spain. *Tapuya: Latin American Science, Technology and Society, 3*(1), 396–414. https://doi.org/10.1080/25729861.2020.1781371
Perler, L., & Schurr, C. (2020). Intimate lives in the global bioeconomy: Reproductive biographies of Mexican egg donors. *Body & Society, 27*, 3–27. https://doi.org/10.1177/1357034X2093632

7
Labeling Ova

Introduction

Selecting bodies able to reproduce desirable health standards is an obligation for clinics according to the Spanish law on assisted reproduction (Law 14/2006). However, the legislation is vague about exactly what clinicians should screen for. More specifically, the law states that donors must be "in good psychophysical health" and that fertility clinics must ensure that donors "do not suffer from genetic or infectious hereditary diseases transmissible to the offspring" (Law 14/2006). However, the questions of what "good" psychophysical health is, what the "transmissible diseases" are, and what counts as a disease all remain unanswered. Rather, the law states that this screening must be done "according to the state of the art of science and the existing techniques at the moment that it is conducted" (Law 14/2006, p. 5). Amid these blurry rules and definitions, in this chapter, I explore how clinicians selected women to take part in egg donation programs. As I show, while clinicians aimed to base their decision on scientific criteria that helped them transform donors into objective stable categories, this process was always contested and shaped by social and cultural norms, as well as commercial ones. As

A. Molas, *Taming Egg Donors*, Health, Technology and Society,
https://doi.org/10.1007/978-981-96-5103-0_7

Wahlberg and Gammeltoft (2018) claim, selective reproductive technologies "are shaped by cultural values and social norms around notions of what 'high quality' or 'good quality'" is (p. 8), rather than by "objective" medical criteria.

The guidelines that regulate this screening were established by the Spanish Fertility Association, which published recommendations for clinics in 2007 (Matorras Weinig & Hernández, 2007). To ensure that the donors are in good psychophysical health, the Spanish Fertility Association states that clinics should screen the donor's blood group and Rhesus factor (Rh); test for Hepatitis B, Hepatitis C, and HIV; and conduct a clinical examination to detect toxoplasmosis, rubella, herpes, and cytomegalovirus. Potential donors should also undergo tests for gonorrhea and chlamydia. In the same document, the Spanish Fertility Association states that clinicians should check that donors and their family have no "genetic diseases or history of cardiovascular diseases, blindness, severe arthritis, childhood diabetes, alcoholism, schizophrenia, depression, epilepsy, Alzheimer's disease, etc." (Matorras Weinig & Hernández, 2007, p. 353).

How imaginaries of social class and race affect notions of the hereditable and transmissible when clinicians try to trace the boundaries between the "biological" and the "social" has been discussed by Laura Perler and myself elsewhere (Molas & Perler, 2020). In this chapter I explore genetic carrier screening and the selection and management of donors' phenotypes. This examination provides an illustrative example of what Sarah Franklin (2013) has termed "technological ambivalence." She uses this concept to describe the contradiction that ARTs generate when, intending to reproduce the "natural" (i.e., normative) order, clinicians need to undertake a number of choreographies that precisely reveal the "unnatural" social norms and the cultural charge they are imbued with. Indeed, while at first sight the aim of IVF might be to produce a pregnancy that results in the birth of a baby, fertility clinic endeavors go beyond combining eggs and sperm in a petri dish, additionally animating a wide range of sociocultural imaginaries about reproduction.

Starting from an understanding of egg cells as "bio-objects" (Holmberg et al., 2011), in this chapter I am concerned with the difficult process of containment and standardization of the genetic and phenotypical condition of egg donors in order to control their eggs' capabilities and future outcomes. This process of bio-identification has the aim of standardizing the genetic and racial past of ova in databases, which inevitably involves a twofold task: assigning new identities to eggs as well as erasing past affective entanglements, so that they can be distributed and commercialized as purified neutral materials.

I first focus on the means, challenges, and consequences of the routinization of genetic carrier screening for egg donors. I show how the entanglements between vague guidelines, the market forces of competition between labs, and the commercial dynamics of the global circulation of eggs led toward specific approaches to reproductive selection. Second, I reveal how egg donors were transformed into phenotype samples for the egg banks of fertility clinics, which tried to provide certain futures to intending parents by controlling and stabilizing donors' racial pasts. I claim that in this context, "phenotype matching" led to a constant negotiation of what "resemblance" meant in a process that was racializing by definition. Finally, I focus on the mechanisms in place to tame the body object. I stress that the body object was managed as a site restricted to medical expertise. In this context, clinicians established an intimate conversation with the body object, from which the biographical subject was excluded.

Genetic Carrier Screening Dynamics

Genetic carrier screening in egg donation was standardized and routinized in Spanish fertility clinics around 2015, as screening became more affordable (field notes). The impact of test results on the selection of egg donors varied across clinics, due to the ambiguity in the current law on assisted reproduction. Below, I discuss the challenges that the normalization of these tests brought to private fertility clinics in Spain.

As explained in more detail in Chap. 2, one of the main innovations of the current Spanish law on assisted reproduction is that it anticipated the proliferation of new techniques due to scientific progress. The law states that the psychophysical condition of the donors should be assessed through "a mandatory protocol of donor screening which will include phenotypical characteristics and the necessary analytical results to prove, *according to the state of the art of science and the existing techniques at the moment that it is conducted*, that donors do not suffer from genetic or infectious hereditary diseases transmissible to the offspring" (Law 14/2006, pp. 5–6 [emphasis added]). The current law leaves the specific screening techniques unnamed, to account for technoscientific advances. In relation to genetic testing, in practice, this means that it is not the fertility clinics, but rather the expanding market of genetic tests that defines what genetic information should be collected.

Genetic carrier screening allows clinics to identify the recessive diseases that an intending donor carries to later use this information to match her with intending fathers or sperm donors. However, this genetic test also entails the possibility of discovering pathologies that might appear later in life or genetic anomalies with unknown health consequences. The implications of this fact will be unpacked later in the section "Rejected donors, future patients?"

The genetic screening protocol announced in the law in 2006, however, was not enacted until mid-2019, when the first guidelines were published by the Spanish Fertility Association (Abellán-García et al., 2019). The document explains basic concepts about genetics and genetic diseases and establishes guidelines for clinics. My fieldwork, however, was conducted in the months before this document was published, so I cannot comment on the degree to which the report may have changed clinical practices.

Instead, in this section, I focus on the different problems that these tests generated for both clinics and egg donors. I first focus on the different understandings of how genetic carrier screening should be used in assisted reproduction. Second, I demonstrate how the new option of shipping eggs abroad led to a stratification of egg markets, in which "clean(er)" eggs became more valuable. Third, I pay attention to the feelings of disorientation shared by many practitioners who felt they lacked the knowledge to make sense of the novelties these tests introduced. And finally, I outline the usually overlooked consequences of the routinization of these tests among egg donors, which resulted in new subjectivities and responsibilities.

Ruling Out Genetic Diseases: Taboos Surrounding Eugenics in Market Dynamics

> It's not about eradicating cystic fibrosis from society, okay? Instead, it's about preventing, if both the man and the donor are carriers, the child from being sick. But if a donor is a carrier but the man isn't, I mean, that child won't be sick. The child will only be a carrier like the donor, and she's healthy, or the man, and he's healthy. I mean, it's about keeping the disease from appearing, not about rejecting this girl because she carries this disease.
> (Virginia, biologist, and egg donation program coordinator)
>
> We are strict in the sense that if she [the donor] is a carrier of any important mutation we don't accept her. And I know that there are other clinics where they accept them and they do the matching with the recipient, and if the recipient doesn't carry congenital deafness, they assign the donor anyway. We wouldn't accept a donor with congenital deafness.
>
> **Anna:** Just in case? Because even if the couple doesn't have it, there are possibilities that the offspring...?
>
> **Aina:** No, no. That first generation would be carriers at most, but for the following, if that baby is a carrier of the mutation and gets together with another person who is a carrier of the mutation, they have a 25% chance of having a sick baby by that mutation, a baby that is born with deafness. Well, we are quite... We do it quite well in this regard.
>
> **Anna:** And what about other more common things such as cystic fibrosis and so on, would that be the same or would it depend?
>
> **Aina:** We would not accept them for sure.
> (Aina, gynecologist)

As the ethnographic examples make clear, amid the dilemmas genetic tests generate, different clinics performed in different ways. While most clinicians stated that the aim of the test was not to eliminate diseases from the population, which many see as a form of eugenics, but to engender "healthy children," others asserted their ethics by precisely being stricter with carriers of highly prevalent diseases such as cystic fibrosis and congenital deafness. Rejecting them would presumably protect not only the present generation of intending parents but also future ones. While the latter is an open statement in favor of genetic selection, the former underpins an interesting rationality in which, although eliminating the disease is unthinkable and unethical, the birth of someone affected by the disease is also unthinkable and unethical. This, therefore, implicitly means that the offspring of those carriers as well as their partners would have to successively undergo genetic carrier screening (and maybe IVF to undergo preimplantation genetic diagnosis [PGD]) to avoid the birth of a potentially affected child in each generation. Like Virginia, other clinicians carefully avoided associations of their practice with eugenics, a sensitive topic in Europe, where it recalls the disastrous consequences of the Second World War (Wade, 2012; Wahlberg & Gammeltoft, 2018). However, as I will discuss throughout, moral dilemmas in genetic selection are resolved instead by market dynamics.

Regardless of Aina's statement, most clinics did not reject donors who carried prevalent genetic diseases, because this would involve ruling out a large number of donors who could be used by carefully matching them with the sperm provider. However, different factors made the stock of "clean(er) egg donors" very valuable. On one hand, while all donors were required to undergo genetic carrier screening, intending fathers had the option to turn down testing, lowering the cost of the treatment. In turn, clinics classified their donors as "apt" (for donors carrying no or very uncommon genetic diseases), "apt with matching" (for carriers of prevalent genetic diseases), or "non-apt" (for carriers of transmissible genetic diseases—usually those that need only one copy of the mutated gene for the offspring to be affected[1]) (field observations).

Some couples chose to save the money and not undergo the test, knowing that in this case, the clinic would assign a "clean(er)" egg donor, that is, a donor without the most prevalent diseases in that population group. In this regard, "clean(er)" donors were more easily marketable.

Another reason for the preference or high value of "clean(er)" egg donors lay in the expansion of egg banks that shipped eggs across Europe. The different regulations regarding donor genetic selection meant that

[1] This pattern of inheritance is called "autosomal dominant." It occurs when one altered copy of the gene (that is, from the mother or the father) is enough for the offspring to be affected by a genetic disorder. In some cases, an affected person inherits the condition from an affected parent. In others, the condition results from a new variant in the gene and occurs in people with no history of the disorder in their family. Some examples of these disorders are Huntington's disease, neurofibromatosis, and Marfan syndrome (MedlinePlus, 2024).

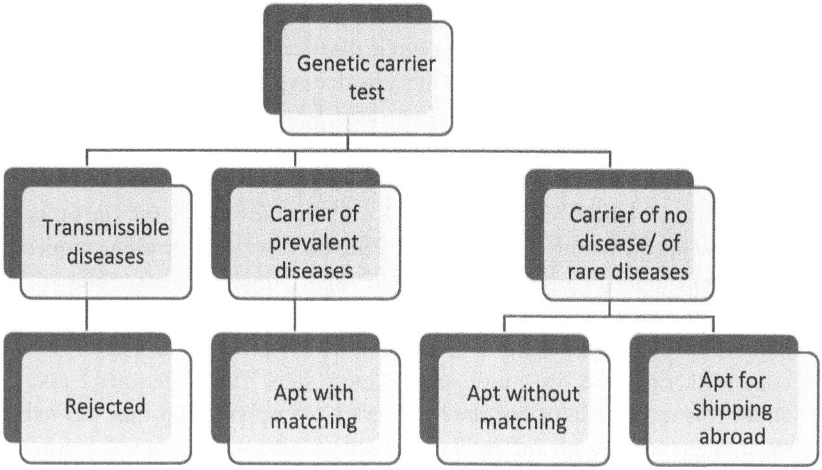

Illustration 7.1 Possible classifications of donors after the results of genetic carrier screening

"clean(er)" eggs were easier to distribute across the continent than eggs that weren't classified as such. That is why I contend that, although the definitions of a "healthy baby" avoided eugenic narratives, local and international market circuits placed higher value on eggs with no prevalent genetic diseases. The following chart (Illustration 7.1) offers a visual representation of the different paths that the eggs of screened donors might take.

Reproduction Under Quality Control

An important point to understand the context of clinicians when they deal with genetic carrier screening is that most of the genetic labs performing these tests are external to the clinic. The clinics are, therefore, customers of these private labs. Criticisms of the logics and practices of genetic labs were raised by clinicians, who reported feeling trapped in both the competition between clinics and between the new genetic tests that continuously popped up. They also argued that the promises that genetic testing gave to intending parents were tricky and even unethical.

The ethical issue is: "Well, where do we stop?" Because nowadays there are between 7000 and 10,000 known genetic diseases. So, do we have to screen for all of them? Because of course, the tests that we do here in Spain analyze around 500 diseases. So you'll look at these 500, but what if she had the 501st? Okay, so we're introducing this false sense of security to the progenitors. [...] We're looking at 0.5% of the possibilities, so we can't guarantee this certainty, right? This is on the one hand. On the other hand, the other debate is: and why only [test] donors? Why don't we do it among couples [who aren't in fertility treatment] as well? Right? To see if they're compatible, and if they aren't, look for a donor. Or if they aren't, then do a preimplantation diagnosis to see if at the end the two abnormal genes have coincided, right? So, we move on one more step, no? Sometimes I joke about this topic and say, "In the end, we'll have a phone app that will tell us our genetic diseases, and we'll check who we do and don't match with," right? Like Tinder, these dating apps, and they'll say, "Well, maybe you're good for a sporadic relationship because reproducing is not really advisable, because we coincide in two or three [genetic diseases]," right? Well, it's a bit of a caricature, but we're moving on with this idea of reproduction being almost under quality control.

(Emilio, biologist and bioethicist)

At the end of the day all of this... since the possibility to do it exists, the possibility is corrupted to turn it into a need. Because down the road, if you have a partner and you want to have a child, you know that there's the possibility of undergoing genetic testing to see if you're compatible. So, maybe in a few years you'll see it as a need because it will be the logical thing to do, to get genetically tested. Then at the end this is a business. And they create this need. There might be cases where this will be great because there will be couples who are not compatible or that have the risk of a disease. So, that couple that was going to have this child naturally, and who had a 25% risk of having a sick child, will end up doing an IVF cycle with a genetic diagnosis and so on. And maybe that child would have been healthy, right? They won't know. Or maybe he would have been sick, and this is fucked up. But maybe she has a healthy baby, and she has hypoxia during labor, and he ends up worse off, you know? That's why I say that in the end you can't control it all, even if it seems like you can. And well, these are possibilities that will become needs.

(Laura, lab director)

The idea of having a quality assurance process for reproduction was criticized but inevitably embraced by some clinicians who, despite promoting genetic carrier screening at the clinic, believed that there were pressing ethical issues to be discussed, as other bigger and more common risks were at play in IVF which were not necessarily acknowledged. Indeed, some clinicians pointed out that the highest risk IVF patients faced was not a genetic disease, but problems during pregnancy and birth due to advanced age, information that some believed should be shared with the intending parents.

> What happens? All genetic tests have improved a ton in these last three, four years, and the prices are going down… very much, which means that with affordable prices you can do lots of tests. Afterwards it's crazy because you never know where the limit is. And people also need to understand that even though you look at many genetic things, no one can tell you that you won't have a problem at birth and that you may have hypoxia and the child… […] You can't make them believe that she will have zero risk of having a child with problems when unfortunately there are things that are not in our hands. We're doing IVF with people who are over 48, who many times have premature births because their womb is old and doesn't work correctly. And all of this needs to be explained well. […] Having so many tests at hand, it's also very important to know the limits of all of this. […] Since there's been this explosion of sequencing techniques and a very big drop in prices, we don't know… we all don't know if we're doing too much, if we're doing too little… I think that nowadays we're all doing too much because we've been caught by… As always, the techniques are faster than the organization and the labeling of things, right? We're all a bit lost, but everyone is very eager for this to be organized, right? We talk about it a lot at conferences.
>
> (Montse, lab director)

Similarly, Gerard (gynecologist) talked about the power that clinicians had when informing intending parents about genetic screening options. According to him, the pressure from private fertility clinics to sell more expensive IVF packages could lead to the manipulation of the intending parents. When I asked him if he believed that couples were well informed when choosing extensive genetic testing for both their donor and the intending father, he said the following:

> You never tell any lies. My view, however, is that the truth can be seen from many different angles. I mean, you as a doctor can manipulate the patient a lot without lying. It depends very much on how you present things, okay? I mean, without lying, I can say… I'll give you an example, I can tell you, "This is a disease that affects almost no one, one in 10,000." Or I can tell you, "*Hostia* [Damn], I'd better do this test because *hostia*, it's one in 10,000, right?" Just because of the way I say it, the answer will be, "Okay, okay, doctor." So just the way I say it will condition you to do it or not to do it, and in either case… In both cases I said the same thing, right?
> (Gerard, gynecologist)

Similarly, Emilio highlighted how private fertility clinics asserted their superiority in relation to the public health system by precisely selling the presumed safety generated by this test:

> There's a dialectic between the public and the private, right? In a way that isn't innocent, if not directly malevolent. What's the message of the private clinics? "Come to us because we'll give you wider screening and you'll have more security that everything will be alright. If you go to the public system, since they don't have money, they won't do this screening and then you'll assume the risks." And we're talking about something that's not banal, right? We're talking about how your children will be, I mean, this is it a very important aspect, right? "Look, if you need 5000 euros, take out a loan." The message is this. […] There's a debate… The law is ambiguous because it says that you have to do all the available tests, so of course, you can do a lot.
> (Emilio, biologist and bioethicist)

The routinization of genetic carrier screening in gamete donation has placed fertility clinics in the uncomfortable position of having to make decisions without clear scientific or ethical guidelines, mainly driven by the strong competition between clinics. In this confusing landscape, some clinicians believed the best way to assert that they were performing ethically was to inform the patients properly about the real benefits and risks of the test, and of IVF more broadly. Whether clinicians did this, however, relied on the ethics of each practitioner, or maybe more accurately, on the politics of each clinic.

Rejected Donors, Future Patients?

As mentioned above, another important aspect of genetic carrier screening is that it can not only reveal the chances of having a child affected by a genetic disease, but also unintentionally discover medical conditions that develop later in life or genetic anomalies with unknown consequences. The problems that such discoveries might generate for the people tested are not unique to genetic carrier screening. Other genetic tests such as Ancestry DNA (Halovic, 2017; Lacaze et al., 2017; Mason, 2017) have generated concerns about the delivery of information before and after a person takes the test, which has been described as particularly problematic as direct-to-costumer genetic testing becomes more accessible (Arribas-Ayllon, 2011; Borry et al., 2011; Savard, 2013). While none of my participants reported having used these testing services, some of them knew that they were carriers of prevalent genetic diseases.

Despite the obvious consequences that the test might have for intending egg donors, clinicians usually did not see this as a problem. Testing women was always seen as an opportunity for egg donors to make the right choices in the future. When a donor was rejected because she was found to be a carrier of a prevalent genetic disease, she was advised to undergo genetic counseling and potentially IVF with PGD should she want to form a family in the future. This was, for example, the view of Pablo (gynecologist and clinic manager):

> The QCarrier [a brand of genetic carrier screening] looks for around 4000 mutations and more than 200 recessive genetic diseases. No woman who wants to have children is doing this. Therefore, I would say that in this moment, donation… Genetically, egg donation is much safer than any other spontaneous pregnancy, this has to be very clear. Every month we have one or several young girls who are carriers of genetic diseases incompatible with donation, and they don't know. And we're doing a big favor to some of them, because some of them are at risk of having children with intellectual neurological deficiency, hematologic [problems], and serious diseases which we rule out when they come to offer their ova. So, they take the genetic information which will be of use for their future children, because we reject them, we tell them that they shouldn't have children

> without consulting beforehand with a gynecologist or a genetic counselor in order to have their own children. Especially diseases tied to the X chromosome, right? Fragile X, which are diseases that have a 25% of probability of having an affected child and so on. Serious illnesses which… and all of this means that egg donation is becoming safer and safer. And without a doubt, safer than a normal pregnancy. Everybody is a carrier of recessive genetic diseases that they don't know about, and that if they don't do a genetic analysis, which now is easy to do, they'll never know. And sometimes there's cystic fibrosis, retinitis pigmentosa, muscular degeneration, fragile X, neurological deficits, things that really… you know? And which will make it so that all women who have been rejected will directly benefit from this information.
>
> (Pablo, gynecologist and clinic manager)

This notion of a "benefit" reinforces the widespread idea of the importance of having a genetically "healthy baby" (Kathryn & Clare, 2010) as well as the patient's responsibility for making this happen. This was also understood in positive terms by several egg donors themselves, who saw in egg donation the opportunity to have information that otherwise they would have never had. Indeed, several donors valued the opportunity to undergo a "very expensive" test for free with valuable information for their future, again evoking the pleasures of temporarily accessing luxurious private health care. This was the case of Stella, who was informed that she had variations associated with two diseases: cystic fibrosis and congenital bilateral absence of the vas deferens, both labeled at high risk of transmission if the sperm provider did not have the genetic test done (1/6452 possibilities of an affected child for the former and 1/295 for the latter).

> They did a blood test to see if I had any genetic diseases which could be… which could be transmitted ehm… I think they analyzed 200 different pathologies or something like that. In fact, I have the documents, which they give you [she shows them to me]. You give consent and they give you the tests which… and really, they are very expensive, the genetic tests. And then I had an interview with the geneticist who explained the tests to me. She said I was quite alright, that I didn't have any out-of-the-ordinary disease that would mean not being able to donate, and she said that if someday I want to have children, the tests could be shown to a geneticist or…

> because they found two pathologies, two diseases that were transmissible. So they would only affect the baby if my partner had them as well. So, they gave me the documents so that I could go someday if I was interested in having children, to see how to do it and solve the problem.
> (Stella, 20 years old, one-time egg donor)

As Stella states, undergoing the test meant for her a new important duty in the future: "solve the problem" by visiting a genetic counselor to have her partner tested, and maybe even undergoing IVF and PGD herself should the partner carry the same genetic disease. A new subjectivity was therefore created among donors who were rejected for genetic reasons or found to carry prevalent genetic diseases: They were women who were "genetically at risk" of giving birth to unhealthy babies, and now that they knew, they should do something about it (see Novas & Rose, 2000). In other words, to put it in Rose's (2007) terms, they became "pre-patients" of the fertility and genetic medicine industry. I contend, therefore, that egg donation not only creates donors but also recruits future patients of these services once they are engaged with the narratives of quality, risk prevention, and responsibility. The violence that such advice might represent in the usual contexts of financial insecurity of egg donors should also be considered.

Moving on to observe how phenotypes were selected, managed, and classified, in the next section, I shed light on how race was constructed and delimited at the clinic.

"What is a Caribbean?": Race in the Making at the Clinic

Nervous and excited in equal parts, I entered the clinic for my first day of observations. My contact and interviewee was Montse, the lab director and assistant manager of the clinic, who after our interview, introduced me to the lab team on duty: two men, Joan and Hector, and a woman, Fina, all seemingly in their mid-30s.

I promised to stay out of their way, and they told me not to worry. Apparently, it was not a busy day. Montse reassured me, "Don't worry, these ones like to talk, you'll see." Joan and Hector informed me that Fina loved to talk about "all these kinds of issues," as, apparently, she liked bioethics. Indeed, Fina told me she was very interested in the debates around surrogacy and altruism. She also wanted to talk about the racial categories they were using to match egg donors and recipients, because maybe I, "as an anthropologist," could help them.

Soon a debate started in the lab after she told me, "Of course, this thing about races is actually not useful," because sometimes they didn't know how to classify people. "What is a Caribbean?" she asked me. Joan joined the discussion, and they argued about the racial category for Egyptians: "Because of course, an Egyptian is Caucasian," said Fina. "No, let's see… it's not really Caucasian, it's more like Arab," Joan replied. "No, but technically they're Caucasians, like Mediterraneans, like us," said Fina. "What do *you* think?" they both asked me.

A few days later I met Gerard for an interview at the public hospital where he worked as a gynecologist, a position he had just started after several years at a private fertility clinic in Barcelona. He told me early in the conversation that he now earned half of what he used to, but he would not change it for anything:

> I find more sense in it now that my work does not depend on the market. I mean, I'm not selling treatments. In the private [clinic] at the end you do your doctor-work as well as you can, and the medicine is done well, but you can't forget that you eat from what you do. So sometimes your medical decisions can be a bit corrupted… It's not in my case, but it's true that at a business level you sometimes do unnecessary treatments. You order unnecessary diagnostic tests that generate more expenses and that are not always necessary for a process. It's a private company.

I told him more about my research and that I loved learning about assisted reproduction because, although it was a high-tech medical field, it reflected many deep social issues. He agreed and told me that actually they could use an anthropologist in their unit as they faced "a lot of ethical dilemmas every day." As an example, he told me that very recently they had a couple from Bolivia "with indigenous features" who needed donor sperm. They could only offer them "Mexican or Colombian sperm," which at the clinic was under the same label, "Amerindians."

However, he continued, for the patients there were differences among these groups, and they did not want sperm from Mexican or Colombian donors. They told the doctors that with those options on the table, they would rather have "Caucasian sperm," which would help with the integration of their child (who was going to live in Spain). He told me that he empathized with the couple and that "if indeed that would help that child, why deny it?" However, another doctor was against it.

Before returning to this story below, I want to emphasize that matching the donor and the recipient was perceived as a challenging part of clinicians' work. Both Gerard and Fina felt that they needed help handling what they perceived as "ethical challenges," which would be more suited to an "anthropologist's expertise" than their own as doctors and biologists in fertility clinics. However, providing "matches" was legally their responsibility. Indeed, the Spanish law on assisted reproduction states that fertility clinics have to provide intending mothers a gamete donor who is as similar as possible to her phenotype (Law 14/2006). But who gets to decide on this matter of resemblance? What categories are prioritized? And given these questions, how do clinicians try to work appropriately and morally? In this section, I delve into these issues to reveal how race was constructed at fertility clinics and how it materialized in egg donor selection practices.

Racist Patients, Moderating Clinicians

As reported by the clinicians that I interviewed at the mid-sized clinics, challenges during the matching process were mainly caused by the pressure placed on them by patients who, apparently, were very concerned about this part of the process and displayed "racist" preferences. Indeed, as I showed earlier in clinicians' narratives about the purposes of genetic carrier screening, the taboos around racism and eugenics remerged around the matching process, provoking feelings of discomfort among the clinicians. The use of the term "phenotypical" (instead of racial) represents an effort to acknowledge and avoid endorsing the racist understandings that led to past violence and discrimination. But in spite of the clinicians themselves, racial categories "become reanimated" in matching practice (Bergmann, 2015, p. 238; see also Russell, 2018). While clinicians aimed to perform objective medical-based matchings, in their practice, they struggled with intending parents to disentangle phenotypical screening and classification from ethnoracial imaginaries. As Thompson (2009) discusses in the context of egg donation in the United States, "perceiving skin tone always involves its intersection with these other attributes and their wider meanings and histories. This means that skin color is an index of legibility whose chromatic properties are deeply relative" (p. 132). The matching process, therefore, serves "as a specific context for the fabrication of race" (Bergmann, 2015, p. 239), which is enacted rather than discovered in the negotiations between patients and doctors and in the screening and selection of egg donors.

While discussing this topic with the clinicians, I usually sensed an urge to make clear not only that they were not racist, but also that they tried to prevent racism from being materialized at the clinic, highlighting their struggles when the recipients made unsettling demands:

> So, what do our foreign egg donation patients want? Look, what they look for… and here their true intentions will become evident, right? But look, it's how it is. They want it to work fast, for the donor to be pretty, and if she can be blond with blue eyes even better, regardless of what the patient looks like. Even if she [the patient] is dark-haired with brown eyes like you and me, all of them want blonds with blue eyes. It's amazing. All Maghrebians suddenly are the only brunettes in their family. "Because in my family everyone is blond," and you say, "Yes sure, the world is full of Berbers, tell me another story," right? The stereotype of the Moroccan is blond with blue eyes and 1.95 [meters], everybody knows! […] I remember an Arab who said, "The most important thing is that she not be Jewish," like, "The donor cannot be Jewish." Excuse me? Do the ova pray? What? "If she is Jewish, I don't want her." And you say, "Sure, you know what? In Spain we don't distinguish for creed, so we don't ask about the religion of the donor, of course." And that was a real conversation, right? "So you don't know if my donor is Jewish?" she said. "Neither yours, nor any donor." I replied. "Well, so how do we do it?" and I said, "We don't do it." I won't ask anyone… In Spain this is an invasion of one's privacy, I won't ask it.
>
> (Ada, international patient coordinator)

Ada did not seem to know, or at least did not recognize as important, that according to Judaism anyone with a Jewish mother is considered

Jewish, regardless of the religion they actually practice. Her perception of this situation as well as examples about Maghrebian patients was that they were nonsensical and racist. Ada's narrative about patients' demands was not an exception. Clinicians at other clinics often highlighted the "racist" demands of patients. In response, clinicians depicted themselves as moderators ultimately aiming to stop this perceived racism from materializing in patient-donor matching. However, and resonating with the results of other studies about matching practices in fertility clinics in Spain and elsewhere in the West (Homanen, 2018; Lafuente-Funes, 2017; Moll, 2019), clinicians especially pointed out their desire to avoid racism when patients not classified as "Caucasian" wanted to be matched with a donor with a lighter phenotype. "In general people are very demanding with the donors. I mean, people want prettier people, taller, blonder, fairer, more fantastic. I mean, everybody wants [someone] like themselves or better" (Gerard, gynecologist).

As Gerard made clear, "better" was always understood as a wish to upgrade to phenotypes associated with higher social status, which might mean different combinations in different regions of the world. Here again, Gerard highlighted the pressure patients put on clinicians to maintain at least their racial status, if not "upgrade" it. Coming back to the story he told me at the beginning of our conversation, I asked him what their final decision was with the couple from Bolivia.

> She was Bolivian, so we classify the donors as Amerindians, like, American indigenous. But of course, for them a Bolivian is not the same as a Colombian because… and it's true, they are quite different. And for them it's offensive because they say, "It's not the same, I want one from my country," so you… and the law says that within the possibilities of that moment you have to offer the most similar donor. So, if at that moment you don't have Bolivians, but you have Mexicans or Colombians, since they're considered the same ethnic group… […]
>
> **Anna:** So then, what happens when he or she expresses their desire to "I'd rather take a Caucasian than a Mexican," what do you do then?
>
> **Gerard:** Strictly, if at that moment you have the option, you should make a match with whatever is most similar to her ethnic group, you should give the same ethnic group, not what she asks for. If she says "No, no, I like Swedish," "No, you're Amerindian, and I have Amerindians. If

> you don't like it, we won't do it." But since they are private companies, many times… you are more flexible and you say, okay. But strictly speaking you should come as close to her phenotype as possible.

Interestingly, "coming as close to her phenotype as possible" in this context represents enabling certain kinds of movements while denying others. In the case presented by Gerard, the couple from Bolivia was directed to Mexican and Colombian donors when Bolivian donors with indigenous features were lacking. However, moving to a donor classified as Caucasian was only considered if they did not have any other Amerindian donors. This means that while the move from Bolivia to Mexico and Colombia was perceived as nearly the same, the move from Bolivia to Sweden was perceived as a means to allow racism and eugenics. "Amerindian" in this context became a social status category rather than a phenotypical one, while whiteness appeared as a resource of power to be protected under the pretext of preventing the racism of those who were more "racially" marginal. Access to this resource, however, was mediated by a private-public dynamic, in which, as Gerard stated, patients from private clinics benefited from more flexibility in how providers met their demands.

Protecting Whiteness and Avoiding Unthinkable Matches

It has been argued that the goal of phenotypical matching is in principle to produce "plausible" babies (Thompson, 2009) for intending parents, or as Peter Wade puts it, children with "race-kinship congruity" (Wade, 2012). In other words, the aim would be to engender children able to "pass" as biological relatives of the intending families (Bergmann, 2012, 2014). This matter is especially salient for heterosexual couples, who have the chance to pass as a "normal" family. The quest for resemblance is therefore a quest for the freedom to keep the donation secret if they wish to, without being confronted by the "resemblance talk" that usually takes place when a baby is born (Becker et al., 2005). Phenotypical congruity with children has also been argued to be a tool for gay men to prevent their children, already facing discrimination at different levels, from being further stigmatized (Smietana & Twine, 2022).

However, resemblance is always understood as something much wider than phenotype. As Joan Bestard (2004) states, through kinship we "relate people to other people and we build the idea of community, we relate different bodies to each other and build the idea of resemblance, we connect ancestors and offspring, and we establish genealogies" (p. 13).[2] Similarly, as Diana Marre and Joan Bestard (2009) put it, kinship is "a way of constructing relations in a network of already existing relatives, a way of placing the new body into the group of the family body and constructing the new individual body as a family member" (p. 65).

Ideas about what is heritable transcend scientific understandings to embrace a range of other imaginaries related to symbolic belonging to a shared past history. As Laura Mamo (2005) puts it, "[t]his construction of relatedness is not based solely on bio-genetic ties nor is it based on social ties. Here, biology and sociality are mobilized to create shared ancestry and the similarities and histories that embodies" (p. 254). An example of the notions that phenotypical matching mobilizes is also visible in Michal Nahman's (2013) examination of how the Israel-Palestinian conflict materializes in the fertility clinic. As she notes, in rejecting the origin of certain eggs and selecting others, "borders and national selves are made and unmade" (p. 127).

According to the clinicians, this intertwining was especially visible when Spanish recipients rejected donors from Latin America. Although in their narratives, these patients told clinicians that they were rejecting donors from certain countries to assure the continuation of their white lineage, I contend that the colonial imaginaries associated with Latin American women also played a role in their rejection. As we will see in the following example, even when the donors were white, the perception of a "Latino" threat remained tangible.

> I have to say that it's curious, and it is a topic to study from the sociological point of view, that Spanish people, and Catalans inside the Spanish people, they don't accept donors from other countries. I mean, non-European

[2] Original text in Spanish, translated by the author.

countries, because I almost don't have European donors. I might have a French donor who is… working here, an Italian, but this isn't common, okay? Donors are either from here or from Latin America, some girls from northern Europe… well northern… from Russia, Ukraine… but I have to say that we don't specialize in these countries, okay? I'm telling you this because there are clinics where you go and all of them are from Russia or from Ukraine. It's not our case, okay? Here basically there are Spanish donors, and when they aren't Spanish, there are many who are from Latin America. Venezuela… Venezuela is a country that's easier for patients to accept, but Spanish people don't, okay? Colombia, Brazil, it depends, because in Brazil there is a lot of diversity, right? Argentina, okay? Chile it depends, okay? Then, what's really hard is Ecuador, Honduras, El Salvador, Venezuela… not to mention the Dominican Republic, okay? Because the Dominican Republic has a lot of *mulata*[3] population, black… so the patients from here they don't accept it. Maybe the African patients, because they know that there are no African donors… […]

Anna: To what extent do you need to tell the intending parents? Let's imagine that you have a donor from Brazil, Argentina, or Ecuador with white ancestors?

Virginia: I tell them. Because of course, she's [the donor] from Ecuador. It can seem to me that she's very white, but I don't know what she has behind that.

(Virginia, biologist and egg donation program coordinator)

As Virginia clearly put it, even if displaying whiteness, the ancestral past of donors from Latin American countries was perceived as a threat to the future of recipient patients. In this context, as Gerard made clear, professionals adopted patients' fears of downgrading their racial status, which led to the precautions they took.

For example, there are clinics who don't want South American donors from some countries. No one wants indigenous South American donors. No one wants Bolivians, no one wants Peruvians. No one. Even the Peruvian recip-

[3] Although the use of the word "mulata" is very widespread in Spain, it is considered derogatory by racialized communities and civil organizations. Other more accepted terms have been proposed, such as *mestizo/a* to refer to biracial people and brown people from Latin America or Latinx descendent (Cárdenas Mena, 2023).

ients, the Bolivians, they want an Argentinian or Venezuelan donor. This is interesting.

Anna: It is! Tell me more about this please.

Gerard: Donors… I mean, for example, a Spanish recipient doesn't want a Latin donor under any circumstance. No matter how amazingly pretty or white-skinned she is. Because… if she has possibilities of having ancestors like grandparents and so on who are very brown, they don't want them. So, you're a Spanish recipient, I give you a donor who is super white, very cute… exactly like you but Argentinean, then they don't want her. Because they say, "Maybe her uncle is from Brazil" and so on, so they're concerned that they will have a black baby, right? These are stupid things, but it's very important. Then ethnic backgrounds, for example, depending on what clinics, if you go [and say], "I want to be a donor" and you are clearly Amerindian Bolivian, like indigenous, they say, "No thanks." […] It's not that they'll be rejected, right? But a lot of times, clinics have a surplus of that because there's less demand. So, in general there's more demand for white donors with light eyes and so on… in this country.

(Gerard, gynecologist)

The meaning of "downgrading" from a Spanish recipient's point of view was well understood by clinicians, who were familiar with (and often implicitly shared) local recipients' racial imaginaries and "affective assemblages" (Kroløkke, 2014) and, therefore, knew what to avoid. But this became more difficult with recipients from other countries and ethnic backgrounds. Indeed, as many clinicians highlighted, working in gamete donation led them to learn about "interracial racism."

There was a patient recently… [laughs] a black woman who told me… I said, "Would you mind if your donor is *mulata*?" [She responded] "Ok, but please, as long as she's not an islander, okay? They're very lazy." [laughs] Of course, then there's interracial racism, not from white to all the rest, which is what we're used to, but that the Japanese don't want Chinese, the Chinese don't want anything that's not Chinese, that the black is like, "Be careful about what country you bring her [the donor] from"… you say *hostias* [damn], of course here we're not used to other types of racism.

(Ada, international patient coordinator)

Of course, the same way we notice the blond and the brunette they notice… the black tone from Nigeria and the black tone from Senegal.

> Which is different, one is darker than the other. And one is taller, and the others are shorter… but we don't know so much. I'm learning that African high lineage people won't accept anything that isn't from their region and so on, so of course, they keep looking because here it's very complicated.
> (Neus, gynecologist)

Interestingly, performing "ethically" in these practices meant to tell the recipients the truth when an exact ethnic-racial match could not be made, which usually involved negotiating about the traits to which the recipients were willing to "downgrade" and which traits to prioritize. This negotiation shows once again the "race in the making" or "racialization" that the matching process inevitably entailed.

> We have a problem especially with mixed race [donors]. Of course. You start to pull out, what about your father? Your grandpa? Your uncle? Your children? Of course, you can't risk it because you think… from both sides. You think, okay, mixed race with what? With an Indian? A Chinese? With a Maghrebi? With a Hindu? I mean… This topic is a bit complicated. To try to see how you do it in order not to fuck it up.
> (Paula, nurse)

Underpinning the argument I have made so far, we see how Paula mixed geographic areas (Maghreb), religions (Hindu), and nationalities (Indian and Chinese) when explaining the difficulties of achieving racial certainty for "mixed-raced" donors. Although again she framed her performance as aiming to prevent racism, this "interracial racism" expertise was required in order not to make matches that would be unthinkable from the patients' point of view.

Race as a Market Driven Assemblage?

Interestingly, while the narratives of patients were labeled as racist, the selection protocols of clinics—designed to satisfy patients—were not. Instead, they were considered to be effects of a society with racial prejudices combined with market dynamics, a definition that evokes the concept of "liberal eugenics" (Braun, 2007; Rose, 2007). With a business approach in mind, clinics focused their efforts on selecting donors who had the greatest possibilities of being matched the patients and therefore generating profits for the clinic.

> At the end of the day, it's a source of income for them, and therefore they want to be donors. The thing is that if we don't have compatible recipients, then to what extent are we interested in carrying out the tests? You know? This happens sometimes, a donor comes in, and you say, "Well, when we have a compatible recipient, we'll let you know and give you the tests," because at the end of the day doing all the donor screening is an economic investment for the clinic, it costs money. If she won't have an outlet, you won't take her.
> (Laura, lab director)

Similarly, Irene (egg donation program coordinator) and Neus (gynecologist and clinic manager) revealed that they had a "sampling" of a diverse range of phenotypes so that they could accept a wide range of patients.

> **Irene:** We have a bank so… if a phenotype walks through the door…
> **Neus:** If an oriental shows up we automatically freeze her. That is, rare phenotypes, a red-haired donor with green eyes. We freeze her automatically, or… a blond with blue eyes who… usually they start the treatment straight away, but if in that moment there's not a blond blue-eyed recipient, we freeze her. We always freeze unusual phenotypes.
> **Irene:** I believe that we have some… some vitrification of all phenotypes.
> **Neus:** A sampling.

The way the doctor put it ("we automatically freeze her") provides a good example of how donors are read through multiple gazes once they are in the clinic. When an intending donor with a valuable phenotype arrived at the clinic, staff immediately pictured her as frozen eggs in the clinic's egg bank, filling a phenotypical gap. Their ability to select and tame egg donors as *biographical subjects* determined whether this goal was achieved or not. Her statement also highlights how the selection criteria appeared to be more flexible with "rare phenotypes." Likewise, Magalí explicitly explained how the phenotypical desirability of donors affected the kind of treatment received in the clinic, reinforcing the findings from Chaps. 3 and 6, where I showed how egg donors' racial biocapital was implicated in the power they would have throughout the process.

> With a lot of donors, we would freeze it all [all the eggs]. Because what happens? That maybe you always have the typical Spanish Caucasians, brunette, a bit like you. White skin, brown hair, brown eyes or… you know? And of course, if the recipient is Swedish with blue eyes and nearly white hair, you are interested in donors like this. So what you do is that when a girl shows up to be informed and she's different from the ones you usually have, you're crazy for her to stay. You give her coffee; you give her whatever she wants [laughs]. It's true. Or for example, you have a recipient who is black. And there are not many black donors, at least here. Then… or B negatives, which are rare blood types. If a girl comes and you say, "Do you know your blood type?" "B negative." We already glance at each other [laughs]. It's like, "This one is coming home with us!" This is what's done.
> (Magalí, biologist)

Similar to what Daisy Deomampo (2019) found in the context of egg donation in the United States, Magalí's statement makes clear that the matching dynamics in fertility clinics not only enacted race, but also conceded capital value to eggs because of their presumed capacity to transmit race. Within this narrative, it is not the clinicians or the clinic that were "racist" but rather the market dynamics on which they depended, which valued certain racial imaginaries and labeled others as surplus or waste.

The obligatory nature of the phenotypical matching between recipients and donors reveals that race is not a stable category and that any ideas around the concept of resemblance will always be contested by different actors across different regions of the world. With the principle of protecting whiteness and preventing patients from "downgrading" to phenotypes associated with lower social status (a notion that is always local and contextual), clinicians found themselves in the position of becoming experts in the racial anxieties of people around the world with the aim of achieving patient satisfaction without crossing "unthinkable" boundaries. The fertility clinic became therefore a site of negotiation, contestation, and ultimately, reproduction of racial tensions.

Taming Bodies Through Check-Ups, "Ecological" Harvesting, and Drugs

If in the previous sections I discussed the problems of trying to assign women and eggs to stable categories, in this section I focus on the way clinicians dealt with egg donors' bodies—that is, with the donor enacted as a body object. The body object was understood at the clinic as a site restricted to medical expertise. Once the biographical subject had been selected and contained, doctors started their intimate conversations with egg donors' bodies in order to achieve a successful egg retrieval. These bodies, however, were not understood as lifeless matter. On the contrary, they were understood as organisms with agency which the doctors had to "get to know" in order to treat them effectively. Indeed, for this taming process, doctors had to discipline the "natural" raw potential of the body through stimulative medication, monitor its reactions through ultrasound check-ups, and contain or correct any imbalance in this tense equilibrium between optimization and risk management (see Molas & Perler, 2020).

In this section, I discuss how the body object became a site restricted to medical experts and to which egg donors as biographical subjects did not have access. I also unpack how clinicians adopted an "ecological" approach to risk management in egg harvesting and the logics and practices behind it. Finally, I focus on a specific example of body containment through the implementation of gonadotropin-releasing hormone (GnRH) antagonist drugs, and how this substance became a tool for optimization, risk management, and relaxation for clinicians in fertility clinics.

Silent Check-Ups

As I mentioned above, Claudia, the gynecologist who did all the ultrasound check-ups that I was able to observe, commonly used words such as *cariño* (roughly, "honey") and *corazón* (roughly, "sweetheart"] to talk to donors when they were in the consultation room. One of the things that struck me the most was that at these check-ups, where the doctor conducted a vaginal ultrasound to check the evolution of the stimulation process, usually happened in silence. When donors asked how it was going, she answered with single words such as "super!" During those observations I saw how, for example, Claudia decided that a donor's stimulation medication should be changed, without explaining to the donor the reason why. I also saw that she found a cyst during an ultrasound and recorded it in the donor's digital record, to be taken it into account during the surgery[4] but didn't tell the donor what she had seen. When I asked her about it, she told me that it was a normal effect of the medication and that it would probably go away in the following days. However, she did not tell me why she didn't offer this explanation to the donor.

Giving no detailed medical information to the donor during the cycle was common practice among the different clinicians I interviewed. The rationale for not doing so was entangled with the realities they found in the clinic among egg donors. When asked, many of them assured me that there was no bad faith in withholding information. The issue seemed to be more related to the presumed uselessness of explaining these technicalities to donors, many of whom, according to them, were basically interested in the money and didn't care about the medical process. In some of their explanations it seemed clear that it was also about not "bugging the donors" (as I described in Chap. 6) with complicated explanations when they perceived that their aim was to get the money as quickly and easily as possible.

> I think that one of the things that you have to care about the most in an egg donation cycle is the donor. Because she's exposed to everything and is suffering everything for nothing. I mean, not for nothing, for the money, but they're not aware that we're giving them all the hormones and so on. I mean, theoretically you explain to her that there is no study to date that proves that these hormones are related to any kind of cancer, you have to explain it a little, but there are a lot of donors who are not even interested in that, I mean, they want what they want so… It's like "I come [to the clinic], and it's already hard for me to come to this check-up [without having to put up with] you explaining nonsense about what the hormones do

[4] It is important to avoid puncturing a cyst during the extraction surgery, which could provoke excessive bleeding.

or don't do or whatever," you know? It depends. It depends a bit on the type of patient—you'll explain a bit more or a bit less how the cycle is going, what you're trying to do, how she'll respond, and so on.
(Eva, gynecologist)

Indeed, the perception of the donor as someone who had a low reproductive health literacy and who was mainly interested in the money was widespread. However, other reasons not related to the comfort of a particular egg donor also played a role. The fear of donors dropping out if the process became tiresome (because of having to engage with too much expert information) or worrying (if they were informed that they had a cyst or that the medication was not having the expected effect) was also a reason why in these encounters clinicians often released as little information as possible.

An "Ecological" Approach to Egg Harvesting

Clinicians also tamed the body object with the drugs themselves, through which they attempted to find an equilibrium between optimization and risk management. Both sides of this equation were equally important and could be translated into financial gain or loss. With this in mind, taking an "ecological approach" was described as desirable by a clinic manager, who made evident the gaze toward the donor as a site of extraction that needed to be cared for with a long-term perspective.

Neus: Let's see, our standard is that the donors do not leave bloated like a balloon and that if she wants, she can donate again. We prefer two donations rather than one donation of 50 ova that will leave you… So, for us, the standard is between 10 and 15, as if it were one IVF. I know that there are clinics where they drug them to get 50 or 30, because you can do it, but these girls can't walk for a week after the donation, if there are no complications… Of course, there are more surgical complications, there are more vascular ovaries, there are more risks of hemorrhage, I mean, you take more risks in everything of course. Obviously, the profit is very high, but we try to make it so that donors leave the same way they entered. I mean, with two ovaries and without much pain. This means not taking them to the limit.

Irene: They all feel quite well when they leave, you see them… The other day, I saw one when she was leaving, and she said, "I didn't expect that I would feel so well."

Neus: This is our philosophy as long as we can stand it. If the egg banks start to [empty]… it's like anything else, right? Fishing for sardines… I don't want to be… you know? But it's like anything else. If we despoil the seas, at the end we won't have fish anymore, so it's a bit the same… In this aspect we are kind of ecological… The donor isn't… the donor is a luxury, right? She's a luxury patient, not an egg machine.

Neus (gynecologist and clinic manager) and Irene (egg donation program coordinator).

Controlling egg donors' bodies, therefore, was not just a risk management measure, but also a project in economic investment in the long term. The "ecological" approach, explicitly described by Neus but also alluded to by professionals at other clinics, meant that having a low-risk policy was also a sustainable way to persuade donors to return. Along with the effort to invest in affective capital that I discussed in Chap. 6, reducing risk was seen as a patient satisfaction measure that should return the investment in the form of a second cycle.

Risk management was mostly done through managing the medication itself. In the following section, I focus on the use of gonadotropin-releasing hormone (GnRH) antagonist drugs. However, other forms of risk management were conducted through contraceptive pills (cycle control by the clinic), antibiotics (administered through pills or as a prophylactic before the surgery to prevent infection), and anxiety pills (prescribed at some centers to be taken the night before the extraction to assure a good night's sleep and an adequate response to the sedation administered on the day of the surgery).

Antagonists

Probably one of the drug developments that resulted in the biggest change for the egg donation industry was the use of antagonists. Scientifically known as Gonadotropin-releasing hormone antagonists (GnRH), these are a class of medications usually administered during the late-follicular phase in ovarian stimulation treatments to prevent a premature rise in serum luteinizing hormone (LH) levels and premature ovulation (Tarlatzis et al., 2006). In practice this means that antagonist drugs can prevent the body from overreacting to the stimulation medication. As several doctors stated, their discovery and implementation resulted in a drastic improvement of the functioning of fertility clinics given their ability to reduce the chances of ovarian hyperstimulation syndrome. According to Montse, this allowed doctors to increase medication at a much lower level of risk:

> Of course, doctors push them [donors] more, they push them more without fear. Before pushing them was, "Oh, I've crossed the line!" I've crossed the line means that she has reacted a lot to this dose that I gave her, right? Yes, I think now doctors they're much more... well, they're much more relaxed. Because that was a constant worry. I mean, I don't even want to talk or think about how we would be in here with the number of donors we have, because we would constantly have donors with low hyperstimulation. It would be constant. Instead, now... we don't even talk about it, right? It's a thing of the past.
> (Montse, lab director)

Indeed, advancements in IVF drugs have reduced one of the most important risks of egg donation, ovarian hyperstimulation syndrome, to imperceptible levels, allowing clinic staff to have a more relaxed approach to their work. In the above quotation, Montse provides an eloquent example of the sociotechnical dimension of the advancements in medicine in IVF. A drug that keeps the body from hyperreacting to the stimulation medication became a tool for clinics to "push them more without fear" and for doctors to be "much more relaxed." GnRH antagonists became in this way, a tool of optimization, risk management, care for donors, and relaxation for doctors in fertility clinics. They exemplify how objects and inanimate matter, in this case substances, can be conceptualized as sources of intimate and caring relationships in their own right and how their careful examination provides a much more accurate picture of the functioning of interactions in health care settings (Ducey, 2010). They also further exemplify how egg donors are handled as body objects to be contained through risk management and an eye on long-term investment return.

Conclusions

Delving deeper into the argument on the ontological duality of egg donors in fertility clinics started in Chap. 6, in this chapter I have focused on the delimitation, selection, and taming practices of the donor as a body object, for which I have referred to the donor enacted as an object of extraction which, nevertheless, had agency that had to be contained.

First, I have explored how genetic carrier screening was managed in fertility clinics. I have argued that the law that required screening according to the current state of science and the techniques available enabled a highly profitable relationship between clinics and genetic labs. The genetic screening of egg donors resulted in the higher valuation of eggs carrying no genetic diseases or non-prevalent ones. I have also examined the consequences for donors when prevalent diseases were found, showing how undergoing this test created new subjectivities and responsibilities.

Second, I have explored how phenotypical matching was navigated at fertility clinics, where I documented the concerns of clinic staff who were torn between the patients' demands and their aim to prevent racism. I have discussed how the search for similarity mandated by law led to certain choreographies while denying others and thus highlighted the protection around whiteness and the ethnoracial imaginaries that were at stake within the Spanish fertility industry. I have argued that phenotypical matching practices converted the fertility clinic into a site of negotiation, contestation, and reproduction of racial tensions.

Finally, I have described the taming techniques in place to contain the body object in order to achieve successful egg donation cycles. Here, I have explained that the bodies of donors were understood as sites restricted to medical expertise. Doctors got to know donors' anatomies and started intimate conversations with them that were not translated to the biographical subject. In this sense, I have discussed the rationalities behind not giving details to donors about the course of the cycle, the "ecological" ethics endorsed to increase the possibility that donors would participate in additional cycles, and the implementation of certain drugs in medication protocols, such as GnRH antagonists, which reduced the chances of hyperstimulation in donors. The use of such medications represented a method of body regulation, a means of optimization, and a source of peace of mind for doctors.

In conclusion, I have highlighted how clinics understood donors as body objects that had to be properly selected and carefully tamed for successful IVF cycles. In doing so, I have focused on the difficult process of containment and standardization of donor eggs' genetic and phenotypical traits that are in place to control their capabilities and future outcomes. The impossibility of standardizing any of these categories has highlighted the social issues that are implicated in selective reproductive technologies. In this context, I have shown how these selection practices ultimately reproduced racial imaginaries and ideas of quality assurance. In the following chapter, I provide a discussion of the main findings and implications of the book in light of the imaginaries about the future of fertility clinics that practitioners evoked.

References

Abellán-García, F., Alamá Faubel, P., Aura Masip, M., Bassas Arnau, L., Castilla Alcalá, J. A., Clúa Obradó, E., De La Fuente, L. A., Guillén Quilez, J. J., Manau Trullâs, D., Rueda, J., Ruiz Jorro, M., & Vendrell Montón, X. (2019). *Cribado genético en donación de gametos*. Sociedad Española de Fertilidad.

Arribas-Ayllon, M. (2011). *Genetic testing: Accounts of autonomy, responsibility, and blame*. Routledge.

Becker, G., Butler, A., & Nachtigall, R. D. (2005). Resemblance talk: A challenge for parents whose children were conceived with donor gametes in the US. *Social Science & Medicine, 61*(6), 1300–1309. https://doi.org/10.1016/j.socscimed.2005.01.018

Bergmann, S. (2012). Resemblance that matters: On transnational anonymized egg donation in two European IVF clinics. In M. Knecht, M. Klotz, & S. Beck (Eds.), *Reproductive technologies as global form: Ethnographies of knowledge, practices, and transnational encounters* (pp. 280–289). Campus Verlag.

Bergmann, S. (2014). "La genética es como la masa de la pizza". El matching y la clasificación del fenotipo como práctica y decisión cultural en las clínicas de reproducción asistida. In E. Pérez Sedeño & E. Ortega Arjonilla (Eds.), *Cartografías del cuerpo: Biopolíticas de la ciencia y la tecnología* (pp. 307–352) Cátedra.

Bergmann, S. (2015). Assisted authenticity: Naturalisation, regulation and the enactment of "race" through donor matching. In V. Kantsa, G. Zanini, &

L. Papadopoulou (Eds.), *(In)Fertile citizens. Anthropological and legal challenges of assisted reproduction technologies* (pp. 231–246). University of Aegean.

Bestard, J. (2004). *Tras la biología: la moralidad del parentesco y las nuevas tecnologías de reproducción.* Publicacions i Edicions de la Universitat de Barcelona.

Borry, P., Henneman, L., Lakeman, P., ten Kate, L. P., Cornel, M. C., & Howard, H. C. (2011). Preconceptional genetic carrier testing and the commercial offer directly-to-consumers. *Hum Reprod, 26*(5), 972–977. https://doi.org/10.1093/humrep/der042

Braun, B. (2007). Biopolitics and the molecularization of life. *Cultural geographies, 14*(1), 6–28. https://doi.org/10.1177/1474474007072817

Cárdenas Mena, Abigal. (2023). *Interculturalidad crítica y antirracismo: Abordando las migraciones y la diversidad con jóvenes.* Fundación Entreculturas. https://redec.org/app/uploads/2023/08/Interculturalidad-cri%CC%81tica-y-Antirracismo-Entreculturas-ES-1.pdf

Deomampo, D. (2019). Racialized commodities: Race and value in human egg donation. *Medical Anthropology, 38*(7), 620–633. https://doi.org/10.1080/01459740.2019.1570188

Ducey, A. (2010). Technologies of caring labor: From objects to affect. In E. Boris & R. Salazar Parreñas (Eds.), *Intimate labors cultures, technologies, and the politics of care* (pp. 18–32). Stanford University Press.

Franklin, S. (2013). *Biological relatives: IVF, stem cells, and the future of kinship.* Duke University Press.

Halovic, S. (2017). The future emerges from the past: Comment on "personal genomic testing, Genetic inheritance, and uncertainty". *Journal of Bioethical Inquiry, 14*(4), 591–592. https://doi.org/10.1007/s11673-017-9818-5

Holmberg, T., Schwennesen, N., & Webster, A. (2011). Bio-objects and the bio-objectification process. *Croat Med J, 52*(6), 740–742. https://doi.org/10.3325/cmj.2011.52.740

Homanen, R. (2018). Reproducing whiteness and enacting kin in the Nordic context of transnational egg donation: Matching donors with cross-border traveller recipients in Finland. *Social Science & Medicine, 203*, 28–34. https://doi.org/10.1016/j.socscimed.2018.03.012

Kathryn, E., & Clare, W. (2010). A "healthy baby": The double imperative of preimplantation genetic diagnosis. *Health (London), 14*(1), 41–56. https://doi.org/10.1177/1363459309347477

Kroløkke, C. (2014). West is best: Affective assemblages and Spanish oocytes. *European Journal Women's Studies, 21*(1), 57–71. https://doi.org/10.1177/1350506813510422

Lacaze, P., Tiller, J., & Ryan, J. (2017). The dangers of direct-to-consumer genetic testing for Alzheimer's disease: Comment on "personal genomic testing, genetic inheritance, and uncertainty". *Journal of Bioethical Inquiry, 14*(4), 585–587. https://doi.org/10.1007/s11673-017-9817-6

Lafuente-Funes, S. (2017). Egg donation in the making: Gender, selection and (in)visibilities in the Spanish bioeconomy of reproduction. In V. Pavone & J. Goven (Eds.), *Bioeconomies* (pp. 253–277). Palgrave Macmillan.

Mamo, L. (2005). Biomedicalizing kinship: Sperm banks and the creation of affinity-ties. *Science as Culture, 14*(3), 237–264. https://doi.org/10.1080/09505430500216833

Marre, D., & Bestard, J. (2009). The family body: Persons, bodies and resemblance. In J. Edwards & C. Salazar (Eds.), *European kinship in the age of biotechnology* (pp. 64–78). Berghahn Books.

Mason, P. H. (2017). Personal genomic testing, genetic inheritance, and uncertainty. *Journal of Bioethical Inquiry, 14*(4), 583–584. https://doi.org/10.1007/s11673-017-9816-7

Matorras Weinig, R., & Hernández, J. (Eds.). (2007). *Estudio y tratamiento de la pareja estéril: Recomendaciones de la Sociedad Española de Fertilidad, con la colaboración de la Asociación Española para el Estudio de la Biología de la Reproducción, de la Asociación Española de Andrología y de la Sociedad Española de Contracepción*. Adalia.

MedlinePlus. (2024, March 31). *Autosomal dominant*. MedlinePlus. https://medlineplus.gov/ency/article/002049.htm#:~:text=Autosomal%20dominant%20is%20one%20of,may%20also%20have%20the%20disease.

Molas, A., and Perler, L. (2020). Selecting women, taming bodies? Body ontologies in egg donation practices in Spain. *: Latin American Science, Technology and Society, 3*(1), 396-414. doi:https://doi.org/10.1080/25729861.2020.1781371

Moll, T. (2019). Making a match: Curating race in South African gamete donation. *Medical Anthropology, 38*(7), 588–602. https://doi.org/10.1080/01459740.2019.1643853

Nahman, M. (2013). *Extractions: An ethnography of reproductive tourism*. Palgrave Macmillan.

Novas, C., & Rose, N. (2000). Genetic risk and the birth of the somatic individual. *Economy and society, 29*(4), 485–513. https://doi.org/10.1080/03085140050174750

Rose, N. S. (2007). *The politics of life itself: Biomedicine, power and subjectivity in the twenty-first century*. Princeton University Press.

Russell, C. (2018). Rights-holders or refugees? Do gay men need reproductive justice? *Reproductive Biomedicine & Society Online, 7*, 131–140. https://doi.org/10.1016/j.rbms.2018.07.001

Savard, J. (2013). Personalised medicine: A critique on the future of health care. *Journal of Bioethical Inquiry, 10*(2), 197–203. https://doi.org/10.1007/s11673-013-9429-8

Smietana, M., & Twine, F. W. (2022). Queer decisions: Racial matching among gay male intended parents. *International Journal of Comparative Sociology, 63*(5–6), 324–344. https://doi.org/10.1177/00207152221102837

Tarlatzis, B. C., Fauser, B. C., Kolibianakis, E. M., Diedrich, K., & Devroey, P. (2006). GnRH antagonists in ovarian stimulation for IVF. *Human Reproduction Update, 12*(4), 333–340. https://doi.org/10.1093/humupd/dml001

Thompson, C. (2009). Skin tone and the persistence of biological race in egg donation for assisted reproduction. In E. Nakano Glenn (Ed.), *Shades of difference: Why skin color matters* (pp. 131–147). Stanford University Press.

Wade, P. (2012). Race, kinship and the ambivalence of identity. In K. Schramm, D. Skinner, & R. Rottenburg (Eds.), *Identity politics and the new genetics: Re/creating categories of difference and belonging* (pp. 79–96). Berghahn Books.

Wahlberg, A., & Gammeltoft, T. M. (2018). *Selective reproduction in the 21st century*. Springer International Publishing.

8

Conclusions

The increasing value of eggs for the growth of the fertility industry worldwide has led to specific systems that automatize their availability, production, and extraction from bioavailable women. Ontologically constructed through different interests and desires, donated eggs are at the same time substances of hope for people seeking to create families, promissory bio-objects for fertility clinics, and a source of temporary economic relief for young women, who, if they meet the clinics' standards, can put their bodily functions at the industry's disposal. The ways in which these desires are orchestrated to operate effectively in the egg donation reproductive market in Spain has been the object of this research.

This ethnography is situated within Spain's highly developed private fertility industry, which performs the highest number of ART cycles in the European Union and is particularly predominant in egg donation cycles. The combination of lenient legislation, mandatory anonymity, and relatively high economic compensation has positioned Spain as a global leader in the egg donation reproductive market.

The main argument of this book is that egg donation is a fragile process that needs a series of careful methods in place for it to work effectively. Egg donation is fragile in the sense that it can fail at any time and,

A. Molas, *Taming Egg Donors*, Health, Technology and Society,
https://doi.org/10.1007/978-981-96-5103-0_8

therefore, it needs to be monitored and controlled throughout all its steps for a successful outcome. The public image of the industry, the selection of egg donors, their behavior throughout the process, the reaction of their bodies, the quality of the extracted eggs, and their proper categorization are all crucial for the functioning of fertility clinics. I have focused on the process through which egg donors and eggs are made operable, asking the following question: How are bodies stabilized for extraction purposes in the Spanish reproductive bioeconomy? As I have argued, these methods for containing women go beyond fertility clinics and operate both at the structural level (where institutional discourses, socioeconomic factors, and regulations converge) and at the micro level of clinical practices (through the relationships established between clinicians, donors, and donors' bodies).

I have conceptualized the control and management of egg donors as a process of *taming*, which has allowed me to offer a new approach to the relations between donors and clinics as well as to move away from a perspective of top-down discipline from clinics to donors. This concept, instead, highlights the more complex interaction between the two, accounting not only for the dynamics of control and coercion but also of care, reciprocity, and mutuality in a highly stratified context. In this book, therefore, I have acknowledged the power imbalance in the clinic while still accounting for the agency of donors' minds, bodies, and eggs—which are perceived by clinicians as agents who contest rules, restrictions, and categorizations. Indeed, egg donors embodied a wild dimension infused with intersecting ideas of gender, age, class, and race that had to be tamed and redirected to achieve successful cycles and more desirable reproduction more broadly—that of the recipient couples. Taming, in this way, emerges as a relational and dynamic process involving a specific kind of emotional, intimate, and embodied labor, as well as an explicitly economic activity. This approach illuminates the tensions and fragility at the heart of the global reproductive market and sheds light on the specific forms of labor and value production that underpin its success. This book has drawn on the points of view and experiences of both egg donors and clinicians. This polyvocal approach has revealed the needs, struggles, and expectations of both and how they were orchestrated to fuel the egg donation bioeconomy in Spain.

After introducing the main topics of the research and the context of the Spanish transnational fertility industry, in Chaps. 1, 2, and 3 I argued that there is an intimate relationship between egg donation and the Spanish labor market, in which egg donors are paradigmatic post-Fordist workers. For most of them, entry to the reproductive bioeconomy was determined by a context of instability and precariousness that expanded across different levels, and where the boundaries between clinical labor and other kinds of precarious labor (inside or at the margins of the labor market) were blurry. Egg donation was in this way a means to achieve their own goals, usually related to securing a better future for themselves in the complicated Spanish labor market defined by short-term contracts, self-employment, risk-taking, and the demand for flexibility.

The legal framing of egg donation as altruistic perpetuated the stigma surrounding the intersection of money and the female body and encouraged women to build narratives around their acts that reaffirmed their gendered care responsibilities. At the same time, this framework generated guilt when donors felt their motivations for the donation did not align with social expectations. The moral weight of this dichotomy between economic interest and altruism was obvious in my interviews with egg donors, who usually came ready to answer the question that they thought would interest me the most: whether they donated for money or out of altruism.

In Chap. 4, I examined how structural factors and clinical practices constrained egg donors to a limited temporality which had implications for their imaginaries about egg donation. I argued that the anonymity framework is key for understanding these dynamics in a particular way: through the legal requirement that donors, recipients, and offspring remain anonymous. Through this framework, not only their identity remained obscured, but also the number of eggs extracted, the number of cycles that the eggs would be used for, their individual value in the global reproductive bioeconomy, and the scarcity of research on long-term risks. These omissions presented donation as an act without enduring consequences and therefore delimited in the present. I also showed how, as reported by women who donated years ago, the intersection of their young age, their financial instability, and their scant cultural capital influenced donors' ability to anticipate and manage risks. As a result, these arrangements tamed donors' imaginaries and separated them from their eggs' value in the reproductive market.

Egg donation is not considered work in official discourse in Spain. For the same reason, the payment received is not labeled as income. In the current law on assisted reproduction (Law 14/2006), it is defined as a donation for which economic compensation is allowed. Clinics adopted this discourse not only when they avoided framing egg donation as a job, but also when they described it as a series of bodily reactions that would happen anyway during a menstrual cycle and which clinics only triggered and further facilitated. Starting from this institutional definition, in Chap. 5, I contested this framework, which ultimately makes invisible the physical and emotional labor of egg donors, by providing a detailed account of what the labor of donorship actually involved for women.

I argued that women had to undergo a number of laborious operations to make themselves bioavailable for the fertility industry. These operations concomitantly carried a subjectification process in which women learned about their biocapital in the industry in relation to their physical traits (in particular height, weight, and phenotype), as well as how to invest it. The arrangements that women needed to make during the cycle included determining how and when to inject the medication, how to juggle the process with their other commitments, and how to manage pain and bodily changes throughout the cycle. These tasks are all part of the unaccounted labor of donorship.

In Chap. 5, I also examined the difficulty women experienced in making sense of their feelings and bodily reactions. The body appeared as a separate entity from the donors. This separation between the subject and the body was connected to the discursive and material practices of the fertility clinics, whose selection and management techniques split the donors into biographical subjects and body objects. The chapter raises questions about the vulnerability of egg donors by examining how the donors used the economic compensation. Most egg donors were not interested in using the economic compensation for the purpose for which it is conceived, such as paying for transportation to the clinic or replacing salary lost because of absences throughout the treatment. Instead, most women squeezed their schedules to the maximum, avoided expenses in their visits to the clinic, and went to work the day after or even the same day of the extraction both to avoid missing days of work (which would result in lost income and the risk of being fired) and to keep the whole sum of money as additional income.

As I showed in Chap. 6, while one would expect that selecting women to become egg donors would be related to selecting bodies that complied with certain health standards and were able to reproduce desirable phenotypes, equally or more important was selecting donors perceived to be responsible and who would obey doctors' instructions throughout the process. Focusing, this time, mostly on the voices of clinicians, I showed how selection and taming practices of egg donors in fertility clinics reveal the way egg donors were enacted as two different entities. On the one hand, they were enacted as biographical subjects (that is, the donor understood in terms of background and behavior) and on the other, as body objects (that is, the donor understood in terms of extractability and of genetic and ethnoracial reproducible capabilities).

The selection of egg donors as biographical subjects was directed to anticipating and controlling the behavior of egg donors throughout the process. Clinicians, and more specifically those at the frontline of the selection process such as administration staff and nurses, described many of their donors as disorganized and untrustworthy. They usually related these traits to their complicated socioeconomic background and their country of origin, invoking imaginaries of class and race. As they explained, their practice was challenging precisely because of the chaos that egg donors generated at the clinic when they didn't follow doctors' instructions. For this reason, clinics had a set of containment techniques to make sure that donors made it until the day of the extraction. While some of these techniques involved threatening them with withholding the economic compensation or with forcing them to reimburse clinics for the cost of the treatment, other techniques were directed to retain donors by providing personalized attention and building intimate relationships that later would create moral obligations and dynamics of reciprocity. As clinicians argued, spoiling and caring for egg donors were perceived as necessary in a context where egg donors had numerous clinics to choose from to undergo the donation. The level of care and patience they were willing to provide was, therefore, related to the level of desirability of each donor.

Finally, in Chap. 7, I examined how egg cells as bio-objects were tamed to control the dangers and ambiguities their existence ex-vivo entailed. With a focus on the genetic carrier screening and the phenotypical (or ethnoracial) screening and selection practices, I observed how clinicians attempted to classify egg donors according to objectifiable categories in order to stabilize their heritable capabilities. I have argued that this process was needed for donated eggs to be neutralized from their pasts so that they could be commercialized as "clean" cells that promised specific futures. Focusing on the routinization of the genetic carrier screening, I observed how, even though the official discourse highlighted an anti-eugenic approach through which intending donors would only be screened to make sure that a particular mutation was not shared with the sperm provider, in practice it was clear that egg donors with no prevalent diseases had more value at the clinics, as their eggs could be assigned to any couple and shipped abroad to other countries with different screening requirements.

Similarly, when examining the management of phenotypical matching at the clinic, I observed that even though clinicians highlighted that they wanted to avoid racist dynamics, they ultimately reproduced racial tensions when attempting to achieve patient satisfaction. This enactment of race happened in the negotiation of what resemblance and difference meant in the context of phenotypical traits. Ethnoracial pasts were therefore contained to assure certain racial futures to intending parents.

In sum, in this book I have shown the carefully tailored methods in place to align systems, bodies, and desires with the interests of the growing reproductive market. By conceptualizing these efforts as a process of taming, I have shown the relational and affective practices of stabilization that underpin the extraction of marketable eggs. Taming, in this context, encompasses a laborious and fragile process of balancing control and care to contain and redirect the perceived extra vitality and unruly biographies of egg donors, which are infused with gender, class, age, and racial ideas. By shedding light on the interplay of power, agency, labor, and stratified reproduction, this book contributes to the scholarship on the political economy of reproduction.

The experiences of clinicians in this endeavor highlight the complications embedded in both the presence of and dependence on human subjects for the supply of raw materials. Professionals' struggles were particularly evident in their hopes and expectations for the future which, they believed, would offer a more comfortable world for reproductive medicine: a world without donors. In the epilogue, I discuss clinicians' imaginaries about the future. Mindful of the fragility of the egg donation process and the challenges of stabilizing eggs and women, clinicians impatiently waited for the time when technoscientific innovation would effectively remove egg donors from the equation of fertility treatments.

9

Epilogue: A World Without Donors

Emilio, a biologist and bioethicist, and I were at his office at the university where he taught when I asked about his thoughts on the debate about the anonymity of egg donation in Spain. Looking a bit bored with my question, he told me that he thought that those arguing for the end of anonymity were genetic essentialists and that this was an old discussion on the table since the latest Spanish law on assisted reproduction appeared in 2006. The debate, however, was about to expire, he believed, and technology would be the way to put an end to it:

> I think this debate has an expiration date. And it will be technology that will solve it, because we won't need donors anymore. In a horizon of maybe… I don't know because I don't have a crystal ball, right? But between 10 and 20 years from now, donors will probably stop being necessary, because the techniques to derive gametes from human pluripotent cells are close to success. Which means that if you don't have ovaries or testicles, it doesn't matter. They'll take your cells from the skin, they'll do an iPSC,[1]

[1] Definition of iPSC in Chap. 2, footnote 6, p. 43.

A. Molas, *Taming Egg Donors*, Health, Technology and Society,
https://doi.org/10.1007/978-981-96-5103-0_9

> they'll make pluripotent stem cells, and with them they'll be able to derive gametes, for example spermatozoids or oocytes, and therefore this debate will be over.
>
> (Emilio, Biologist and Bioethicist)

Professionals at different clinics invoked hopefully the idea that donors would no longer be necessary in the near future. In this epilogue, I argue that clinicians' hopes and desires for the future reveal their present worries and anxieties. To put an end to these concerns they only need to wait for further technoscientific advances, which should provide the definitive solution, removing donors from the process.

As I have examined throughout this book, dealing with donors was experienced as a tiresome burden, for which many clinicians did not feel sufficiently trained. The practices of hiring specific people (often without medical training) to deal with donors and segregating recipients and donors into different spaces to hide donors' markers of class and maintain the professional image of the clinic reveal the ambivalent relationship between donors and clinics: Donors were both necessary and troublesome. Among the systems in place to ensure that egg donors would bring the donation to fruition, I have revealed how egg donors were also converted into different bodies that were selected and tamed through a series of specific sociotechnical methods. The elevated price of the drugs and tests they underwent, and the possibility that donors could simply "disappear" in the middle of the process, made clinicians dependent on them. But most importantly, donors not only required a large investment of material and structural resources from the clinics, but also brought with them multiple ethical dilemmas that threatened the growth of the reproductive industry. From time to time, the media have raised questions about the anonymity of donation, the delay in the rollout of the national donor registry, and the potential exploitation of egg donors (Bassa & Gonzalo, 2023; Gómez, 2019; Remacha, 2019; Virgili, 2020; Zafra, 2019).

For Emilio, the solution to the problem of donor anonymity, for example, will not come through politics or policy. The solution is technoscientific and merely requires patience. In this way, important issues affecting egg donors and the population more broadly were perceived as "liminal," for which there is little cause for debate. The fantasized future landscape looked much more sophisticated. Such fantasies evoked a moment when clinics would finally be only for intended parents and when professionals would be able to exercise their medical expertise fully rather than having to deal with and care for bothersome donors with complicated biographies.

An Expiring System: The (Much Awaited) End of Egg Donation

But what does the *near* future look like? I asked my participants. According to them, everything seems to indicate that the market for eggs will expand further. In this sense, egg banks are expected to grow in number, and the practice of shipping eggs abroad to become much more routinized and common. Some clinicians also highlighted the increasing popularity of social egg freezing—that is, the freezing of one's own eggs for non-medical reasons. However, most of them were not very optimistic about its potential to solve the problems of egg availability. As Laura (lab director) claimed, the problem was that the motivation to undergo social egg freezing usually came at a time when women were at the limits of their oocyte quality.

> The girls who are 20 and who freeze their eggs are a minority, right? The number of 20-year-olds who think, "Maybe I'll want to have children in the future, and maybe I won't be able to," they are very few, very few. They come from 35 onwards. So, it has already passed. Genetically and at the oocyte level you're already late. So I think that society is pushing in this direction when instead it should push to make it easier for women to have children earlier.
>
> (Laura, lab director)

The "improvement" of genetic tests, the legalization of gene editing of embryos as well as less invasive PGD screening, which would bring "more genetic safety" (Aina, gynecologist), were also believed to be issues that would deeply transform assisted reproductive medicine in the short-term.

Other clinicians also pointed out that Spain would soon have to come to terms with surrogacy. The debate on legalizing surrogacy in Spain has been on the public and political agenda for several years now, with parliamentary groups sustaining diametrically opposing views (Boussarie, 2021; Noriega, 2019). Several clinicians who participated in this research were in favor of legalizing surrogacy. Nevertheless, as Emilio and others stated, a much more promising and less ethically challenging future awaits in 20 to 50 years: a future without donors (and without surrogates as well).

Egg Donation as "the prehistory of genetics"

> Well, and now they're going to… and you'll see with this, ha! This thing of the artificial wombs and so on. This, and also, they're trying to create pluripotent cells in order to make eggs from them. Directly, right? I mean, you take a cell from anywhere that is already specific. You take it back so it can be pluripotent, and you push it to be an egg. So, you don't depend on the ovarian reserve, I mean, from the ovaries of that woman. Instead, you'll be able to create an egg of hers from another cell. Well, this is where all this thing of reproduction will go at the end.
>
> (Eva, gynecologist)

According to some clinicians, like Eva, the rules of the game are about to change. The new game board promises to finally get rid of third-party IVF participants, and with them, a big part of the current public and bioethical debates.

One of the most interesting conversations that I had on the topic was with Pablo, gynecologist and manager of a middle-sized clinic in Catalonia. After a first day conducting observations in the lab, I had the chance to interview him. Enthusiastic about sharing his views on the history of assisted reproduction, its current moment, and its future, he talked with me for almost two hours. Like Emilio, Pablo evoked a hopeful future where the "hack job" of replacing the eggs of one woman with the eggs of another would come to an end. In that future, egg donation would be seen as "the prehistory of genetics."

> The thing is that I think that donation will last a few more years, I don't know if 10, 15, 20, 30, but there will be a moment in which with stem cells we probably won't need eggs to generate an embryo, but it will be these same stem cells that will generate the egg so that we can create an embryo without donation. [...] We won't need an egg bank because our fertility clinics will have transformed, right? There will be... no doubt there will be eggs, but generated not by a donor but by a lab, right? Therefore, we won't need a woman to donate eggs anymore. [...] Imagine all this cell manipulation used in reproduction, right? This is the future of fertility clinics in 30, 40, or 50 years. In which we'll take a cell of yours, even if you're menopausal, we'll transform it into an egg, okay? We'll fertilize it, maybe with sperm that does not come from a sperm bank, but from a male cell that has been transformed, and from that, those eggs transformed from a stem cell and with that sperm transformed from a stem cell, which will have not come from a testicle or an ovary, we will make an embryo. This is the future, okay? And we won't need an egg bank or a sperm bank. Instead, fertility clinics will be genetic banks that will transform cells. [...] Therefore, I will transmit your genetic information, which is what people want. What you want is to perpetuate your genetics. Now we're in this in-between [phase] where what we're doing is a hack job, okay? By using the genetics of another person because we're not able to transform the genetics of one cell to another one. [...] The people living in 100, 200, 300 years, they'll say, "These people used to take eggs of other women to get another one pregnant." This will be obsolete. This will be the prehistory of genetics.
>
> (Pablo, gynecologist and clinic manager)

According to Pablo's futuristic perspective, the present is already an obsolete past. The present becomes a time that is dead or "posthumous," following Marina Garcés' (2017) conceptualization of the "posthumous condition." As Garcés states, this concept refers to "the rupture of the eternal present and the beginning of a no-time" (p. 23). "The posthumous condition is the afterwards of a death that is not our real death, but a historical death caused by the dominant discourse of our time" (p. 26).[2] Transposing her argument to my fieldwork, I contend that the present is understood at the clinic as a deferral after a death that has already been announced for years at conferences and in papers. This perspective poses

[2] Original book in Spanish. Fragment translated to English by the author.

important problems when it comes to opening debates, as the imaginary of a unique linear time makes waiting for the future—rather than taking political action—seem like a plausible option.

There is another fantasy tightly connected to this one: Not only will egg donors disappear from the schema, but also surrogacy will be obsolete. Clinicians believed that advances in the artificial womb would end that discussion as well. In the last part of my interview with Pablo, Montse (lab director) joined us and joked about how Pablo loved to talk about these topics. "I'm sure no one so far gave you his perspective," she laughed. "The fertility clinics will be genetic clinics, right? Of cell cultivation and reproduction, okay? So, the same way that you sell eggs, somehow you will sell cells…" continued Pablo, at which point Montse intervened with a half-smile correcting his language, "You *distribute* eggs, you *distribute* them…."

Pablo: So, we can all be eunuchs, we can castrate them all, right? And all menopausal, but they will still have children. Because moreover, what's the issue? Surrogate mothers will exist, so you won't need to have…

Montse: The womb, they already have the womb… There is already an artificial womb. Nowadays there is a prototype of… […]

Pablo: All of this is on YouTube; I'll look it up for you.

Montse: It's very creepy, I warn you. Because it's like a closed vacuum bag with the little one inside. So, the problem of gestation before the lungs are developed is basically a breathing problem. Until the end of the pregnancy the lungs are not mature enough. So, what is totally unfeasible with children born at 20 weeks is basically a breathing issue, right Pablo?

Pablo: Yes, yes. [Still looking for the video on the Internet]

Montse: So, if [the fetus] doesn't breathe properly it can't develop everything else. So, this system is… they put them in a closed vacuum bag…

Pablo: I'm telling you about this because from the anthropological point of view, I like it…

Anna: Of course.

Pablo: [with the video ready to show on his computer screen] You'll freak out, okay? This is the experiment they did with a lamb, okay?

Montse: But this can be done… This is already happening, right?

Pablo: Well, but this is a first thing, but we're talking about 50 years, right? Or 25.

Montse: No, for this you are talking much less, maybe 5 years, right? It's very developed.

Pablo: Yeah, yeah, but this is an experimental artificial womb, this is a lamb, there are many things but…

Montse: So, it's a pre-incubator.

[We start to watch the video as we keep talking. Image 9.1]

Pablo: I remember when I was working at a public hospital in Barcelona, they started to talk about… There were some comments, about uterine transplants, okay? And getting her pregnant afterwards. There are children born like that already. I mean, you're born without a womb, you're compatible with your mom, they take out your mom's womb, they put it in you, you get pregnant, you have children, and the same day or the next day, they take it out again. If you had told me 25 years ago that there would be uterine transplants that would result in pregnancies I would have said, "Come on, this is impossible," because when you do a uterine transplant, you have to give her anti-rejection medication. This anti-rejection medication is usually incompatible with the fetus.

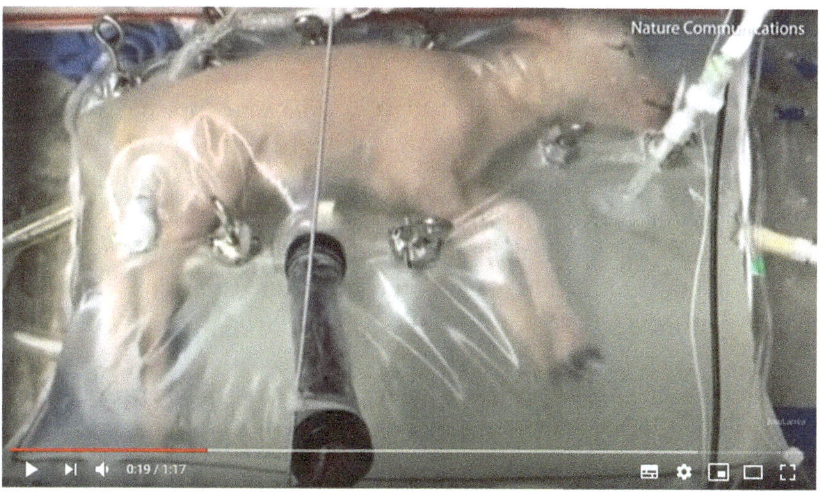

Image 9.1 Frame of the YouTube video: "Desarrollan un útero artificial" [An artificial womb is developed]

Montse: Well, and chemo that they're doing now on pregnant women which don't affect the little one…

Pablo: This seemed impossible but… of course. […] like fetal surgery, okay? You take it out or you do fetoscopy. You operate on it, and you put it back inside, you stitch it up, and the pregnancy continues. This seemed impossible 50 years ago. And we're doing it. What they will do in reproduction in 50 years, I tell you, we'll start from the cellular level, but we'll go much further. […]

Montse: It's very creepy. [watching the image of the lamb in the artificial womb on the screen]

Pablo: I freaked out, and I've been a gynecologist for 30 years.

Montse: To me this image that looks like they're in a closed vacuum bag gives me the creeps… then when I look at it from the other side I think, "*Hostia* [damn it], but this is what is needed, right?"

Pablo: I can have a bag, I take the artery, the umbilical veins, I connect them to a machine that is outside the body, which oxygenates the blood, and I stick in liquid, I stick in nutrients, and that guy lives outside… that fetus, 4, 5, 6 weeks. Of course, if they told you that 30 years ago you would have said, "This is impossible," right? But no, nothing is impossible, I tell you that.

The mix of fear and excitement that Pablo and Montse expressed in our conversation was common among other clinicians. While they had been practitioners at these clinics for a long time, they were passive witnesses of the new technoscientific advances that were presented annually at the conferences they attended. And while they emphasized the possibilities that these advances could bring, such as the end of ethical dilemmas surrounding the use of third-party participants, they also highlighted the drawbacks of these techniques. Laura (lab director), who was pregnant at the time of the interview, pointed to the repercussions this could have for business.

So, I think, the day that artificial wombs are developed, reproduction as we know it will change. All the obstetric complications won't exist. I wouldn't like it, okay? I think that… the thought that a machine will give birth for me… I wouldn't like it. I understand that humankind at the end needs to

> go through a lot of things, good and bad experiences in the life cycle. But if you think about it from the business point of view for example, there wouldn't be maternity leave, women wouldn't have problems… Of course, this would make the eyes of businessmen twinkle. I mean that… I don't know, I mean, you can do so many things… Biotechnology also allows this. They generate hybrids between biological tissue and nanotechnology, you know? At the end… Well, I don't even know.
>
> (Laura, lab director)

Fertility clinics inhabit an imagined and fantasized future, which is clearer, brighter, and better. The present is portrayed as a "hack job," an obsolete past waiting for the technoscientific future. Therefore, debates on the liminal practices undertaken in this already dead time, such as egg donation, lack interest and relevance.

Applying iPSCs in fertility medicine resolves the problem of genetic reproduction and is expected to facilitate fertility clinics' work by ending the participation of donors. As in many other industries, technoscientific innovations progressively make it possible to manage without first-line employees, without providing solutions to the precarious context of these flexible workers situated in a post-Fordist context. If the iPSCs prophecy comes true, the question remains: What other niches will appear for women who, like the participants of this research, look for means to secure a better future?

The linear and unique version of the future carries with it a deterministic logic in which action in the present is seen as irrelevant, as if the trajectory of events were already tangible and immutable. In opposition to this discourse, I refuse to think about this book as a historical report, documenting the remains of a zombie system waiting for the light of a new day to come. On the contrary, I argue that confronting this version of the story requires being able to grasp the logics and rationalities of the present to open space for further debates: What will it mean to reproduce in the future? Through which discourses will the ideas of health and safety be built? Who will mediate these increasingly technological conceptions and births? This space can lead us to create niches in this dead present to imagine other futures, or even better, other presents.

References

Bassa, M., & Gonzalo, À. (2023). *Gens anònims* 3Cat. https://www.3cat.cat/3cat/gens-anonims/video/6207094/

Boussarie, P. (2021, June 25). Ciudadanos busca perfil social abanderando la gestación subrogada. *El Mundo*. https://www.elmundo.es/espana/2021/06/25/60d609aa21efa0d5468b4592.html

Garcés, M. (2017). *Nueva ilustración radical*. Editorial Anagrama.

Gómez, B. (2019, October 29). Las donantes de óvulos se quejan de falta de cuidados frente a los cambios físicos y psicológicos del proceso. *elDiario.es*. https://www.eldiario.es/nidos/donantes-ovulos-quejan-falta-cuidados_1_1237444.html

Noriega, D. (2019, August 6). El debate comienza con el nombre: ¿vientres de alquiler, gestación subrogada o por sustitución? *elDiario.es*. https://www.eldiario.es/sociedad/vientres-alquiler-gestacion-subrogada-comienza_1_1403458.html

Remacha, B. (2019, February 6). El Comité de Bioética Español determina que las donaciones de espermatozoides y óvulos no deberían ser anónimas. *elDiario.es*. https://www.eldiario.es/sociedad/comite-bioetica-determina-donaciones-espermatozoides_1_1144997.html

Virgili, M. (2020, March 8). El cos de la dona, la fàbrica del segle XXI In *Les dones i els dies*. https://www.ccma.cat/catradio/alacarta/les-dones-i-els-dies/el-cos-de-la-dona-la-fabrica-del-segle-xxi/audio/1065984/

Zafra, I. (2019, July 21). España, el granero europeo de óvulos. *El País*. https://elpais.com/sociedad/2019/07/19/actualidad/1563531405_529791.html

Appendix: Overview of Egg Donor Participants

Illustration A1 (Intending) egg donor participants' summary table

Name	Age at the interview (Age at the first donation)	Country of origin	Donation cycles
Alba	25 (19)	Spain	1
Alexia	21 (20)	Spain	1
Andrea	29 (27)	Spain	2
Ángela	22 (20)	Spain	2
Arlet	27 (21)	Spain	1
Berta	21 (20)	Spain	2
Diana	24 (22)	Spain	1
Emily	23 (23)	Ecuador	2
Helena	27 (20)	Spain	2
Júlia	34 (27)	Spain	2
Laia	26 (21)	Spain	2
Lídia	27 (20)	Spain	1
Lucía	22 (18)	Uruguay	4
Maria	34 (32)	Spain	1
Martina	22 (20)	Spain	4
Merche	26 (19)	Spain	1
Mica	22 (22)	Uruguay	2
Mireia	28 (22)	Spain	1
Nancy	22	Peru	0

(*continued*)

A. Molas, *Taming Egg Donors*, Health, Technology and Society,
https://doi.org/10.1007/978-981-96-5103-0

(continued)

Name	Age at the interview (Age at the first donation)	Country of origin	Donation cycles
Noemí	22	Spain	0
Rocío	28 (19)	Spain	1
Sara	25 (19)	Spain	2
Sonia	31 (21)	Spain	1
Stella	20 (18)	Argentina	1
Valentina	21	Colombia	0

The manufacturer's authorised representative in the EU is Springer Nature Customer Service Centre GmbH, Europaplatz 3, 69115 Heidelberg, Germany. If you have any concerns regarding our products, please contact ProductSafety@springernature.com

Printed and bound by CPI Group (UK) Ltd, Croydon, CR0 4YY
30/09/2026
02225254-0006